Advances in Structure and Activity Relationship of Coumarin Derivatives

Advances in Structure and Activity Relationship of Coumarin Derivatives

Advances in Structure and Activity Relationship of Coumarin Derivatives

Edited by

Santhosh Penta
Department of Chemistry,
National Institute of Technology, Raipur, Chhattisgarh, India

AMSTERDAM • BOSTON • HEIDELBERG • LONDON
NEW YORK • OXFORD • PARIS • SAN DIEGO
SAN FRANCISCO • SINGAPORE • SYDNEY • TOKYO
Academic Press is an imprint of Elsevier

Academic Press is an imprint of Elsevier
125, London Wall, EC2Y 5AS.
525 B Street, Suite 1800, San Diego, CA 92101-4495, USA
225 Wyman Street, Waltham, MA 02451, USA
The Boulevard, Langford Lane, Kidlington, Oxford OX5 1GB, UK

Notices
Knowledge and best practice in this field are constantly changing. As new research and experience broaden
our understanding, changes in research methods or professional practices, may become necessary.

Practitioners and researchers must always rely on their own experience and knowledge in evaluating and using
any information or methods described herein. In using such information or methods they should be mindful of
their own safety and the safety of others, including parties for whom they have a professional responsibility.

To the fullest extent of the law, neither the Publisher nor the authors, contributors, or editors, assume any
liability for any injury and/or damage to persons or property as a matter of products liability, negligence or
otherwise, or from any use or operation of any methods, products, instructions, or ideas contained in the
material herein.

ISBN: 978-0-12-803797-3

British Library Cataloguing-in-Publication Data
A catalogue record for this book is available from the British Library

Library of Congress Cataloging-in-Publication Data
A catalog record for this book is available from the Library of Congress

For Information on all Academic Press publications
visit our website at http://store.elsevier.com/

Working together
to grow libraries in
developing countries

www.elsevier.com • www.bookaid.org

CONTENTS

LIST OF CONTRIBUTORS

Santhosh Penta
Department of Chemistry, National Institute of Technology, Raipur,
Chhattisgarh, India

V. Rajeswer Rao
Department of Chemistry, National Institute of Technology, Warangal, India

Gudala Satish
Department of Chemistry, National Institute of Technology, Raipur, Chhattisgarh,
India

Archi Sharma
Department of Chemistry, National Institute of Technology, Raipur, Chhattisgarh,
India

CHAPTER *1*

Introduction to Coumarin and SAR

Santhosh Penta
Department of Chemistry, National Institute of Technology, Raipur, Chhattisgarh, India

1.1 WHAT IS COUMARIN?

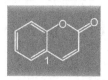

The coumarin[1] (benzopyran-2-one, or chromen-2-one) ring system [1], present in natural products such as tonka beans, warfarin, and clover leaf, displays interesting pharmacological properties. The parent molecule was first isolated by Vogel[2] from tonka beans. The coumarin ring can be looked upon as arising out of a fusion of a pyrone ring with a benzene nucleus. The derivatives of coumarin usually occur as secondary metabolites present in seeds, roots, and leaves of many plant species.[3] More than 300 coumarins have been identified from natural sources, especially from green plants. These varying substances have disparate pharmacological, biochemical, and therapeutic applications.[4] As for physical properties, coumarin is a white crystalline, volatile compound. It smells like vanilla and has a melting point of 341–344 K.

1.2 TYPES OF COUMARIN AND EXAMPLES

There are various ways to classify coumarins according to their chemical structure, occurrence, or synthesis, thus it creates several classes of coumarins. Here we mention roughly categorized coumarins on the basis of their structure[5] given in Table 1.1:

a. Simple coumarins
b. Furanocoumarins
c. Pyranocoumarins
d. Pyrone-substituted coumarin

1.3 WHAT IS SAR?

There is an appropriate way for interaction between drugs and the respective biological site to achieve effective biological activity. A higher level of information regarding the molecular level of a drug and its mechanism of biological activity leads to better understanding for developing the drug with optimum efficiency. These structure-related properties of a drug can be resolved, either through computational

Table 1.1 Classification of Coumarins

Classification	Features	Examples
Simple coumarins	Hydroxylated, alkylated, alkoxylated on coumarin	Hydroxy Coumarin
Furanocoumarins	Five-membered furan ring attached to benzene ring	Psoralene
Pyranocoumarins	Six-membered pyrone ring attached to benzene ring	Seselin
Pyrone-substituted coumarin	Substitution on pyrone ring	Bishydroxy coumarin-Dicounmarol

method (*in silico*) or experimental method with *in vivo* and *in vitro* conditions. Structure-activity relationship (SAR) is an approach that is designed to find relationships between chemical structure of a ligand and biological target of studied compounds. Likewise, for all molecules it is also hypothesized that similar molecules have similar activities. This assumption is well considered as the guiding principle of SAR.

1.4 WHY SAR?

The traditional methods of drug design are now modified by concise methods of SAR, which is found to be highly influential in the drug design process. The geometrical and electrostatic interaction is responsible for causing the effective biological activity. Geometrical interaction involves the complete three-dimensional spatial arrangement of target site and ligands, whereas the electrostatic interaction involves the electronic effect, hydrophobicity, solubility, and so on. However, a number of these interactions cannot be characterized in suitable numerical considerations to describe several attributes. Thus, the mechanism of chemical interaction between the ligand and targeted biological active site needs abundant information. SAR exists to explain the ways of interaction between the ligand and receptor and is applicable in optimization of ligands for better interactive behavior for highly specific and potent bioactive drugs.

1.5 APPLICATION OF SAR

SARs are the traditional practices of medicinal chemistry via which we try to modify the effect or the potency (i.e., activity) of bioactive chemical compounds by modifying their chemical structure.

Medicinal chemists use the techniques of chemical synthesis to insert new chemical groups into the biomedical compound and test the modifications for their biological effects. This enables the identification and determination of the chemical groups responsible for evoking a target biological effect in the organism. Figure 1.1 shows the several applications of studying the SAR.

1.5.1 Pharmacokinetics Studies Via SAR

Pharmacokinetics involves four basic steps: absorption, distribution, metabolism, and excretion (ADME). Bioavailability of a drug, defined as the amount of drug that is actually available at the active site of receptor, is an important parameter of pharmacokinetics. Bioavailability of a drug mainly depends on two major steps:

1. Absorption
2. Metabolism

Absorption depends significantly on the extent of aqueous solubility and its lipophilicity, which can be stabilized by adding an alcoholic group, an acidic group or carboxylic group, or the like in the lipophilic core moiety. In most drugs, metabolism reduces

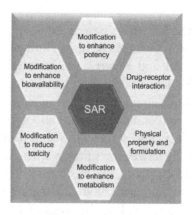

Figure 1.1 Application of SAR.

bioavailability. SAR is helpful in determining the solubility, lipophilicity, rate of reaction in metabolism, metabolites, toxicology, and interaction between drugs.

1.5.2 Drug-Receptor Interaction Studies Using SAR

Most drugs display extremely great relationship of structure besides specificity to have highly potent pharmacological activity. The interaction of drugs occurs with active sites of a biological target that has specific three-dimensional structure and protein-like properties. Drug-receptor interaction requires a minimum of three points of attachment.

The binding between the drug and its receptor occurs by either of these two ways:

1. *Reversible binding.* Weak ionic bond formation takes place between the drug and receptor, such as hydrogen bonding or van der Waals bonding.
2. *Irreversible binding.* Strong bond formation takes place, as in a covalent bond.

Several *in silico* methods were developed to determine SAR for the interaction between the target receptor site and drugs, with the illustration of types of binding. They reveal the extent of binding affinity for the set of drugs.

1.5.3 Chemical, Physical Property and Formulation

SAR has arisen as an important tool for the understanding and development of chemical and physical properties and the formulation of drugs. SAR provides insight into the important aspects of molecular structure, which is effective for the potent activity. Such information may assist in devising an organized approach to drug design of lead compounds with more desired and specific potency.

1.5.4 Drug Modification

Today SAR provides several *in silico* methods developed for optimization and development of drugs. These include the statistical method, validation method, quantum analysis, and artificial networks modeling, among others.

1.5.5 Toxicity Studies Using SAR

The dose of a drug must be accurate. If the administration of a drug is too much, it may cause toxicity; if it is too low, it will not show any activity. SAR is applicable in determining the minimum effective concentration of drugs. Specificity of drugs is also one of the most important factors of SAR studies. If a drug is not specific for its receptor or biological target, it will definitely cause side effects.

1.6 WHEN SAR STUDIES ARE DONE

Drug design involves a gradual process, starting from a compound having interesting bioactivity and ending with its optimized consequence. In drug design, SAR imports a qualitative association of biological or chemical activity of a compound. Designing the drugs mainly involves two methods: protein-based drug design and ligand-based drug design. Protein-based drug design depends completely on the crystallized structure of protein. But the fact is, crystallization of a protein is still challenging. So, ligand-based drug design is more persuasive over protein-based drug design. SAR technique is mainly used in ligand-based drug design for the optimization of a drug molecule.

1.6.1 Screening

The evaluation or investigation of something as part of a chemical structure survey is known as *screening*.

1.6.2 Lead Molecule

A *lead compound* in drug discovery is a chemical compound that has pharmacological activity and whose chemical structure is used as a starting point for chemical modifications, in order to improve potency selectivity or pharmacokinetic parameters.

1.6.3 Pruning

Pruning is the refinement of lead structure. It is done to determine the pharmacophore.

1.6.4 Pharmacophore

A pharmacophore is a spatial arrangement of functional groups essential for biological activity. Thus, pharmacophores define the geometrical property of the ligand.

1.6.5 Mechanism of Interaction with Target Site

The specific site of a receptor or a grid of receptors acts as target site or binding site. The pharmacophore binds at the active site by means of electrostatic bonds, such hydrogen bonds, as well as by covalent bonds. The lead compound changes its own conformation to attain the maximum interaction with the active site.

1.6.6 Structure-Activity Relationships

The conformation of the lead compound as well as its binding parts will provide information regarding the mechanism of binding a ligand with its respective target site. In SAR, correlation studies are preferred. On the basis of relative studies of interaction the scores of the ligand are defined. The ligand having the highest correlation with the biological target is taken for the analysis before entering Phase 1.

1.6.7 Lead Optimization

Optimization of a lead compound is very important to obtain the specified and potent drug. SAR is very useful in determination the ways for attending the maximum potency of a drug with very high specificity, since the interaction between the biological compound and the ligand gives the appropriate information for required geometrical and electrostatic properties.

1.7 NEEDS OF SAR

The following are requirements for SAR:

• SAR is used to determine the parts of the structure of the lead compound that are responsible for both its beneficial biological activity, that is, its pharmacophore, and its unwanted side effects. Therefore, SAR is needed to determine the physic-chemical behavior and biological activity in coherent ways.
• SAR is used to develop new drugs that have increased activity.
• SAR allows researchers to identify the changes in pharmacological properties by performing minor changes in the drug molecule.
• SAR allows researchers to understand and explain the mechanisms of activity within a set of ligands.
• SAR allows researchers to save on product development costs. Estimation may well decrease the necessity for prolonged and costly animal tests. Lessening the use of animals reduces their pain and uneasiness.

1.8 SAR OF COUMARIN IN DISEASE BIOLOGY

Coumarins display a remarkable array of biochemical and pharmacological activity. Certain members of this group of compounds may significantly affect the function of various mammalian cellular systems. Coumarin is an aromatic compound that has a bicyclic structure with lactone carbonyl groups. The presence of an electronegative atom is effective for hydrogen bond formation and for solubility, to some extent and aromatic ring is responsible for having hydrophobicity. These phenomena are the cause of better interaction of the molecule with a receptor site. The substitution of coumarins makes them more significant for effective bioactivity. Numerous types of coumarins have been synthesized and also are present in nature. With different structures due to the various types of substitutions or pharmacophore in their basic nuclei, they are significant in showing effective and diverse classes of biological activity.

Based on the substitution pattern, coumarins show *anticancer,*[6,7] *anti-HIV,*[7,8] *anticoagulant,*[9] *antimicrobial,*[10,11] *antioxidant,*[12] *hepatoprotective,*[13] *antithrombotic,*[14] *antituberculosis, antiviral, anticarcinigenic, and anti-inflammatory* activities.[15]

The pharmacological and biochemical properties and therapeutic applications of simple coumarins depend on the pattern of substitution in basic coumarin moiety. Therefore, there is a need to conduct a careful study of the SAR of coumarins.

1.9 SUMMARY

Structure-activity relationships, or SARs, deal with the influence of the functional groups present in a drug on its biological activity. SARs determine the pattern of influence employed in the drug's design and in the synthesis of many drugs of desired pharmacological activity. Coumarin is an important chemical compound with versatile activity. Thus we can conclude that studying the SAR of coumarin will provide prominent information on synthesis of specified and effective drugs.

CHAPTER 2

Antimicrobial Agents

Santhosh Penta

Department of Chemistry, National Institute of Technology, Raipur, Chhattisgarh, India

2.1 INTRODUCTION

Microbial diseases have been widely distributed in the world in each era. Several microbes are capable of generating resistivity to antimicrobial agents.[16] This fact has caused researchers to generate new and modified medicines. Antimicrobials are broadly used against bacteria, viruses, and fungi. Herein we discuss the SARs of respective antimicrobial agents.

2.2 WHAT IS AN ANTIBACTERIAL AGENT?

A substance derived from microbes or artificially synthesized that either inhibits the growth of or destroys bacteria is known as an *antibacterial agent*. If this agent completely kills the bacteria, it is known as being *bactericidal* (Figure 2.1), and if it inhibits the growth of bacteria, it is known as *bacteriostatic*. Antibacterial compounds are given in concentrations that are nontoxic or harmless to the host and can be

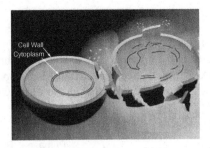

Figure 2.1 Bactericidal activity. www.youtube.com/watch?v=qBdYnRhdWcQ.

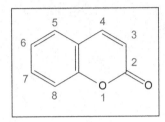

Figure 2.2 Numbering of Coumarin nucleus.

used as effective therapeutic compounds to stop bacterial diseases. Antibacterial activity of a number of different compounds was compared at concentrations resulting in maximal effect on the bacterial cell at which growth was arrested. Coumarin-based antibacterial agents were selected to study through SAR; these illustrate strong influence in activity with the change in type of substituents along with the position on the coumarin nucleus. Thus, effects of several functional groups and their presence at particular positions of the coumarin nucleus are elucidated as the numbering given in Figure 2.2.

2.2.1 Functional Group at C_3 Position
2.2.1.1 Alkyl Chain
Carboxylic Group with Alkyl Chain
The carboxylic group attached to the third position of the coumarin ring through an alkyl linker has been found to influence the anti-helicase effect. This signifies the favorable interactions of carboxylic acid with the bacterial DNA helicase. In alkyl chains, the presence of a methylene (CH_2) group [2] in between the coumarin nucleus and the carboxylic group revealed three- to four times less potency compared to an ethylene (CH_2CH_2) group [3] against DNA helicases; see Figure 2.3 of *B. anthracis* and *S. aureus*. Compounds having propylene ($CH_2CH_2CH_2$) group [4] and [5] exhibited potency similar to that of the compound [3].[17]

During replication (copying of DNA), DNA helicase creates a replication fork to unwind the DNA strands to allow the formation of daughter strands from the parental template. DNA helicase inhibitors inhibit the unwinding of DNA stands; hence replication also stops. Thus, bacterial helicase inhibitors are applicable to cease bacterial growth.[18]

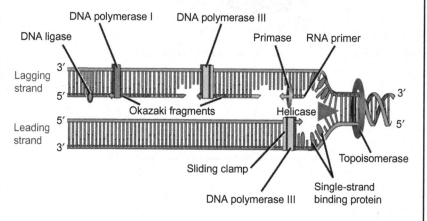

Figure 2.3 Replication of Bacterial DNA (Function of Helicase). www.vanderbilt.edu/vicb/ DiscoveriesArchives/dnareplication_ipond.html.

2

3

4

5

Where, R =

Amide Group with Alkyl Chain

The amide compound [6], substituted with tertiary amine, exhibited helicase inhibitory activity better than a compound having the carboxylic group [3]. Whereas, neutral amide derivatives [7-8] were not active for helicase of *B. anthracis* and showed the significant decrease in contradiction of helicase of *S. aureus*. This was expected due to the presence of a (+) charged amine group along with the linkage moiety, so it might spread in part having solvent or a hydrophilic region to create extra association with enzymes, a cause for increase in affinity.[17]

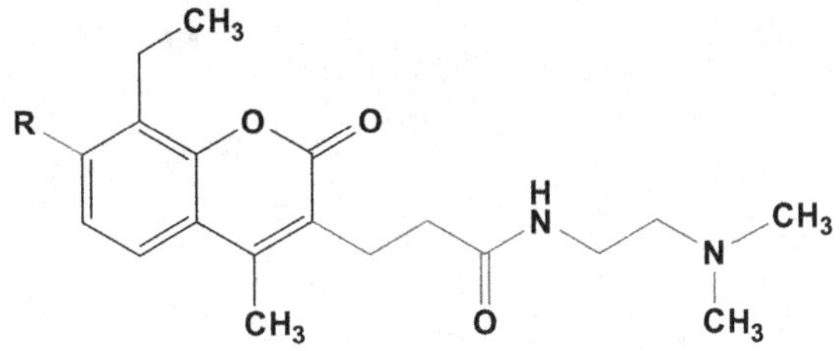

6

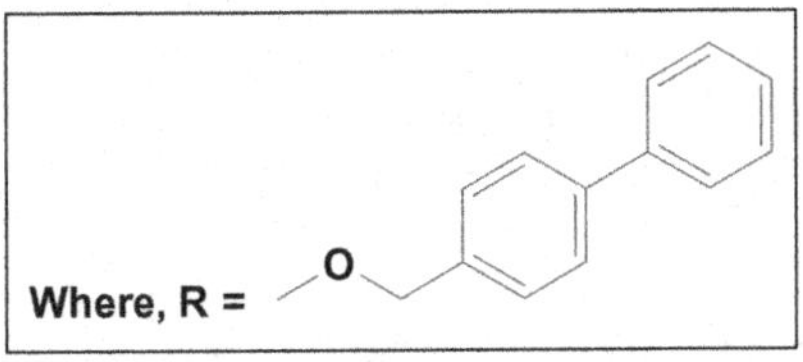

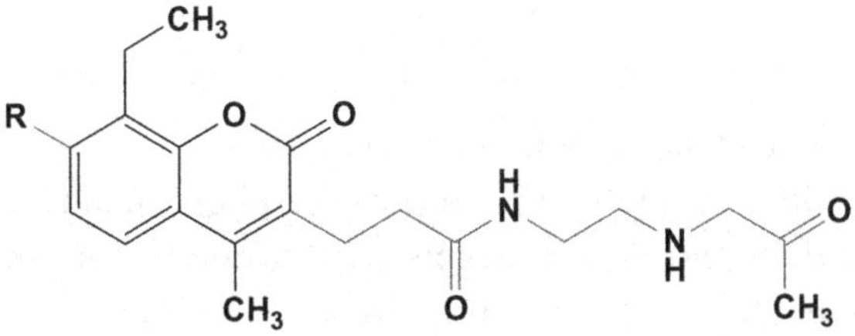

Where, R =

7

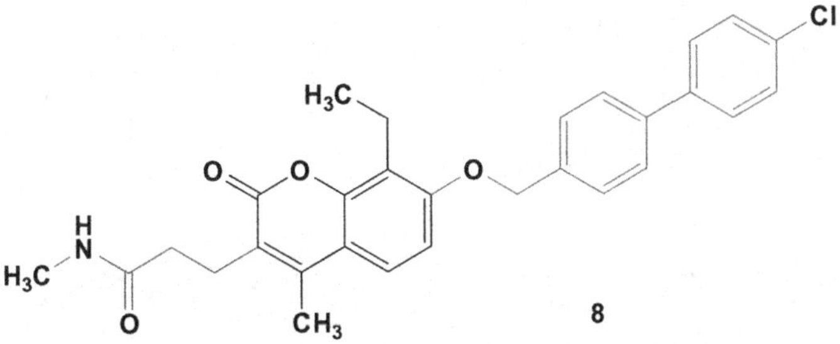

8

2.2.1.2 Carbonyl and Alkoxyimino Group

Coumarins substituted with 3-ketone [9], 3-ester [10], and 3-alkoxyimino [11] derivatives were found to be extremely active against the gram-positive strains, together with *Staphylococci* and *Enterococcus* species.[19]

2.2.1.3 Amide and Hydroxamate Groups

Hydrogen bonding is one of the important phenomena for interaction of drugs with their biological targets. An amide group and its derivatives can play an important role in hydrogen bonding. Hence coumarins substituted with 3-amides [12], its derivatives [13], and hydroxamate derivatives [14] displayed greatest antibacterial activity, mostly against novobiocin-resistant strains.[19]

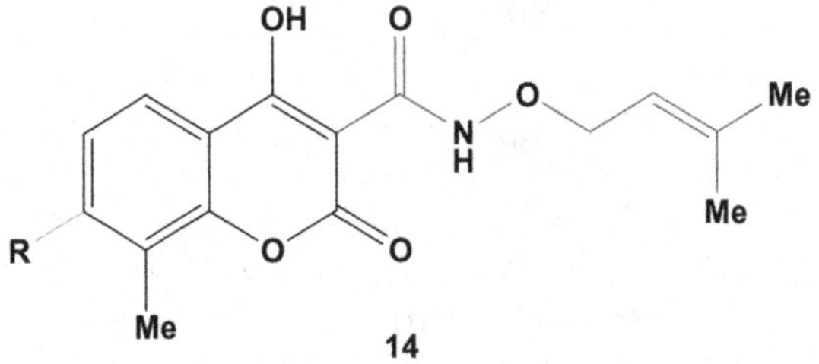

14

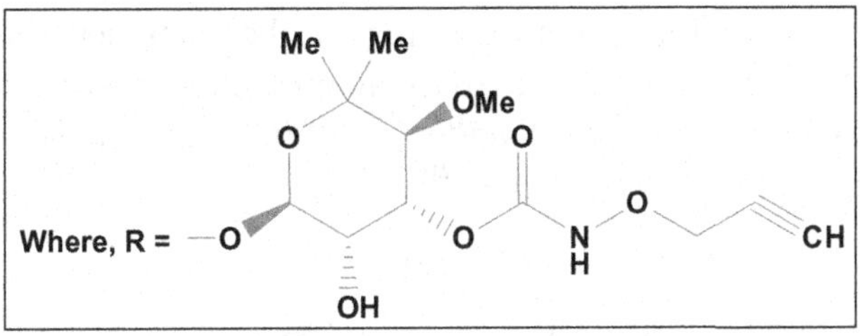

Where, R =

2.2.1.4 Benzyl Group with Different Substitution

The benzene rings substituted at methylene-bis position were estimated for their efficiency as antibacterial agents, mainly to counter gram-positive and gram-negative bacteria such as *Bacillus subtilis*, *Staphylococcus aureus*, *Escherichia coli*, and *Klebsiella sp.*, respectively. [20]

t-Butyl Group

The tertiary butyl group substituted at *para* position [15] showed the lowest antibacterial activity among the benzene phenyl-methylene-bis (coumarin).[20]

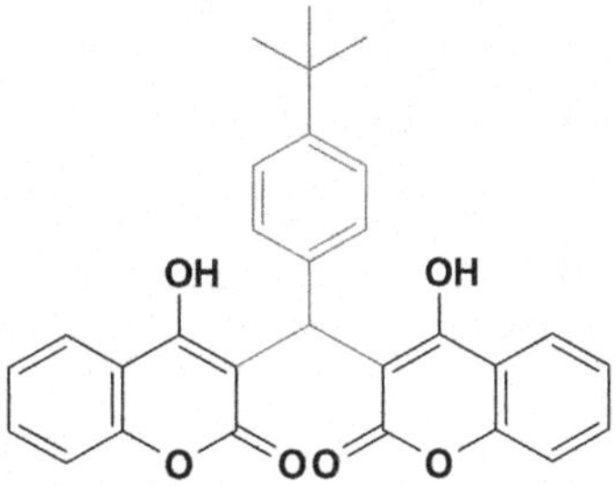

15

Hydroxyl Group

The benzene ring substituted with hydroxyl group at the methylene-bis position [16] displayed the greatest antibacterial activity.[20]

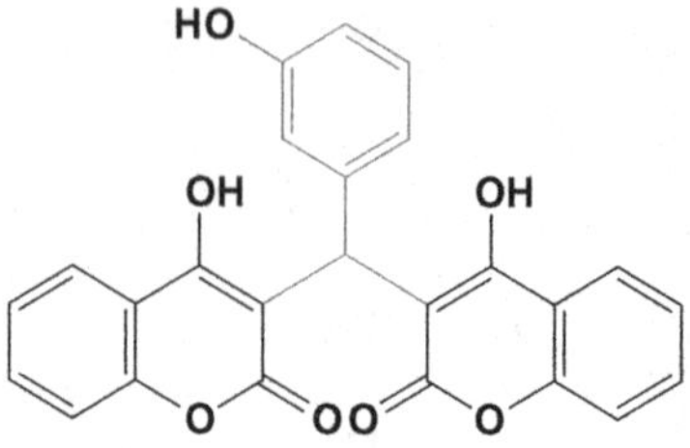

16

Methoxy Group

The substituted methoxy benzene ring at the methylene-bis position
[17] had the lowest inhibitory activity against gram-positive and gram-
negative bacteria such as *Bacillus subtilis, Staphylococcus aureus,
Klebsiella sp.*, and *Escherichia coli.*[20]

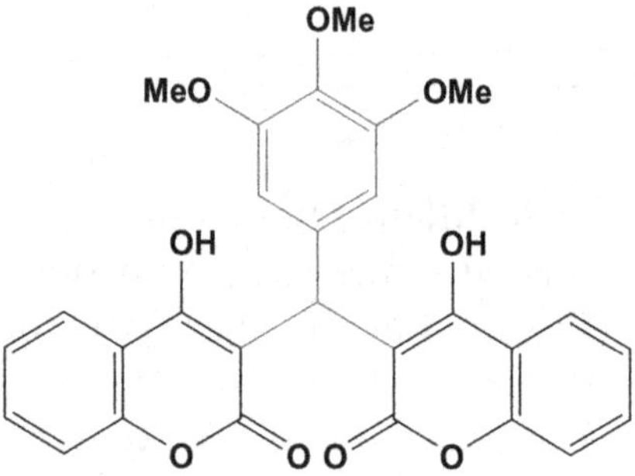

17

Halide Group

The benzene rings with halide-substituted derivatives [18a-e] had
effective antibacterial activities in comparison to the dicoumarol.[20]

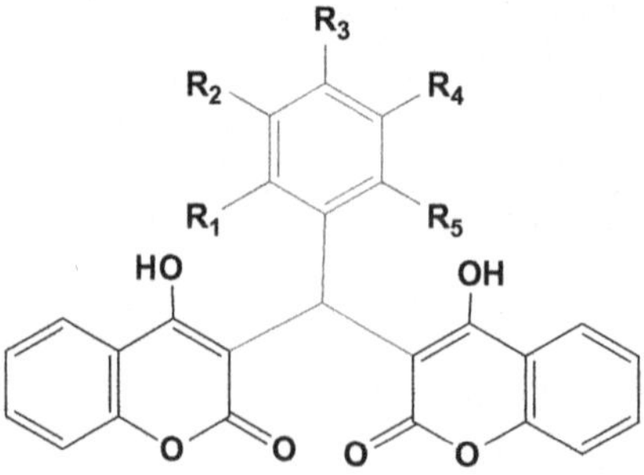

18

18	R_1	R_2	R_3	R_4	R_5
a	Cl	H	Cl	H	H
b	H	Cl	Cl	H	H
c	Cl	H	H	H	H
d	H	H	Cl	H	H
e	H	Cl	H	H	H

2.2.2 Functional Group at C_4 Position

2.2.2.1 Small Aliphatic Group (CH_3)

The C-4 methyl-substituted antimicrobial compound [19] that exhibited the strongest inhibitory effect against *H. pylori* is 16 times more potent than the lead compounds (psoralen, isopsoralen, and xanthotoxin).[21]

19

As compared to unsubstituted methyl or ethyl-substituted derivatives, it was revealed that methyl-substituted compounds [3] possess potent inhibitory activity for *Bacillus anthracis* and *Staphylococcus aureus*. Absence of methyl groups from the C-4 position [20] of coumarin moiety intensely lessened the potency.[17]

20

3

Where, R =

2.2.2.2 Iodinated Aryloxymethyl Coumarin

Iodinated-4-aryloxymethylcoumarins were screened for two mycobacterial strains, as shown in Figure 2.4 (*Mycobacterium tuberculosis* H37 RV and *Mycobacterium phlei*).[22]

Mycobacterium, genus *M. phlei*, is a fast-growing mycobacterium. Patients diseased from *M. phlei* normally get healthy when treated with antibacterial agents, whereas the genus *M. tuberculosis* is the main causative agent of tuberculosis.[23]

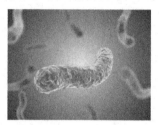

Figure 2.4 Mycobacterium tuberculosis. http://drugdiscovery.com/upimages/1379626198_M.tuberculosis.jpg.

2-Idophenol Group

Halogens substituted on coumarin at the sixth and seventh positions of compound [22(h-j)] displayed effective activity, whereas the mono-methyl compounds [22(a-b)] showed least activity. The rest of the derivatives demonstrated moderate inhibitory activity.[22]

a = 6-CH$_3$
b = 7-CH$_3$
c = 7,8-Dimethyl
d = 6-OCH$_3$
e = 7-OCH$_3$
f = 5,6-Benzo
g = 7,8-Benzo
h = 6-Cl
i = 7-Cl
j = 6-Br

22

3-Idophenol Group

Bromo and chloro compounds [23h-j] exhibited a potent increase in activity against *Mycobacterium tuberculosis* H37 RV and *Mycobacterium phlei*. The dimethyl and benzo compounds [23c, g-h] exhibited moderate activity in this series. Likewise, the mono methyl compounds [23a-b] were moderate active. The change in the position of iodine (second to third position) was unaffected on activity of methoxy compounds.[22]

a = 6-CH₃
b = 7-CH₃
c = 7,8-Dimethyl
d = 6-OCH₃
e = 7-OCH₃
f = 5,6-Benzo
g = 7,8-Benzo
h = 6-Cl
i = 7-Cl
j = 6-Br

23

4-Idophenol Group

For *Mycobacterium tuberculosis*, inhibitory activity increased for compounds bearing chlorine [24h-i], but the activity remains the same for the bromo compound [24j]. The chloro compounds showed excellent activity with MIC 3.125 mg/mL and were more potent than standard drugs streptomycin and equally potent with pyrazinamide. The methyl compounds [24a-b] exhibited moderate activity and were unaffected. Dimethyl and methoxy compounds [24c-e] also showed moderate activity. The lease activity was exhibited by benzo compounds [24f-g]. The moderate inhibitory activity against *Mycobacterium phlei* was showed by chloro and bromo compound [24h-j].[22]

a = 6-CH₃
b = 7-CH₃
c = 7,8-Dimethyl
d = 6-OCH₃
e = 7-OCH₃
f = 5,6-Benzo
g = 7,8-Benzo
h = 6-Cl
i = 7-Cl
j = 6-Br

24

2.2.2.3 Sulphur Derivatives

Among sulphur-substituted 2-thioimidazo derivatives [25, 26] and 3-thiotriazo derivatives [27], 3-thiotriazo derivatives were found to be the most favorable antibacterial agents.[24]

2.2.3 Functional Group at C_5 Position
2.2.3.1 Amine Group

The antimicrobial activities against *H. pylori* were radically decreased when the nitro group at C-5 was reduced to an amine, or the methoxy at C-8 [28] was demethylated to a hydroxyl group [29].[21]

2.2.3.2 Nitro Group

Compound [30] with a nitro group at C-5 and methoxy at C-8 showed fourfold stronger against *H. pylori* than the lead compounds (psoralen, isopsoralen, and xanthotoxin).[21]

30

2.2.4 Functional Group at C₇ Position

2.2.4.1 Alkenyl and Alkynyl Group

Compounds [31] and [32], having an alkenyl or alkynyl substituent respectively at the C-7 position of the coumarin ring, were found to be inactive compounds for helicase.[17]

31

32

2.2.4.2 Carbamoyl Group

Adverse supercoiling of DNA gyrase was inhibited by carbamoyl derivatives [33], as similar to novobiocin. A decrease in the inhibitory activity was observed in *N*-alkyl derivatives. *N*-propargyloxycarbamate was developed as an effective 5-methylpyrrole-2-carboxylate bioisostere, which can improve the *in vitro* inhibitory activity for bacteria and negative supercoiling of DNA gyrase.[19]

33a R= -NH-Allyl
33b R= -NH-Propargyl
33c R= -NHiPr

33

In carbamoyl derivatives, increasing the length of the substituent more than a 3-carbon chain [34] can cause a decrease in inhibitory potency.[19]

34

2.2.4.3 Complex Alkoxy Group
Electron-Withdrawing Group

As in comparing with unsubstituted derivatives [2] and substituted derivatives, a compound having substitution mainly with an electron withdrawing group such as F, Cl, CF_3, and cyanide on a phenyl ring were shown increased potency by two fold. The chloro compound [5] was the most potent among biphenyl derivatives. This compound revealed the activity (IC50 = 1.0 μM) to counter *S. aureus* helicase and equivalent value against *B. anthracis* helicase with the IC50 value of 1.5 μM.[17] Substitution of a halide group, mainly F, and the Cl atom at the C-6 position, ortho with respect to the alkylamino group at C-7 in the quinolone series, enhanced the potency of derivatives.[24]

5

2.2.4.4 Phenyl Group

Compound [35], having a phenyl ring at the C-7 position of coumarin moiety, showed a weak inhibition effect for helicases.[17]

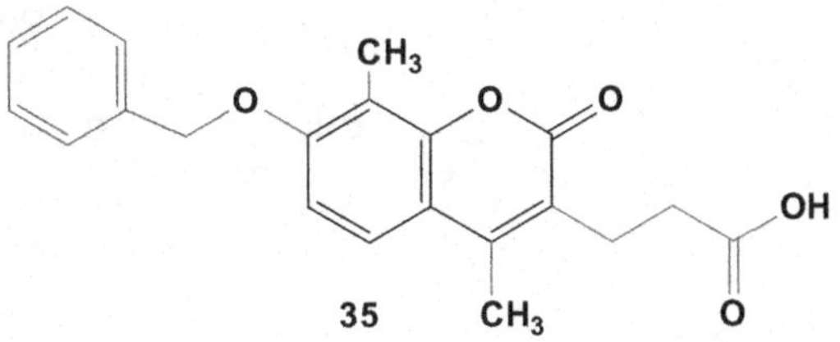

35

2.2.4.5 Biphenyl Group

Compounds with a biphenyl group at the C-7 position of the coumarin nucleus were shown effective inhibitory activity against DNA helicase. The 1,4-biphenyl substituted derivative provides the greatest potency against both *S. aureus* and *B. anthracis helicases*. 1,4-biphenyl [38] vs. 1,2-biphenyl [36] and 1,3-biphenyl compound [37]; similarly, 1,4-biphenyl [41] vs. 1,2-biphenyl [39] and 1,3-biphenyl compound [40].[17]

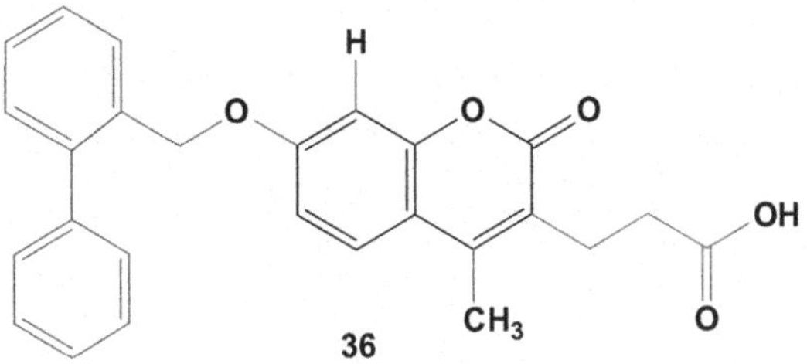

36

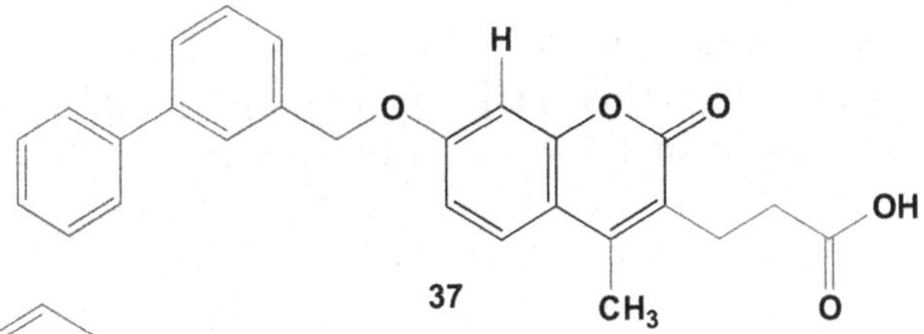

37

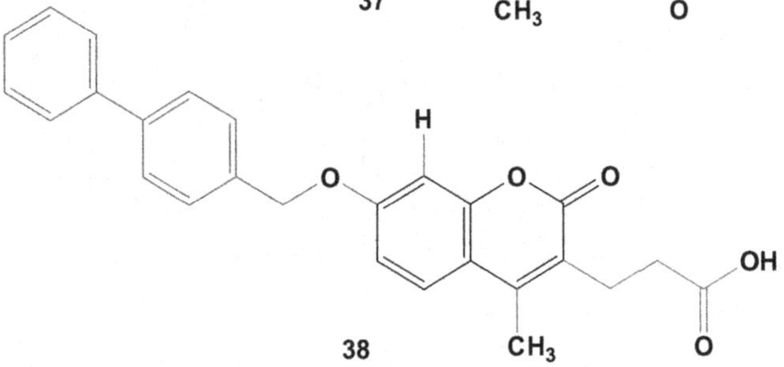

38

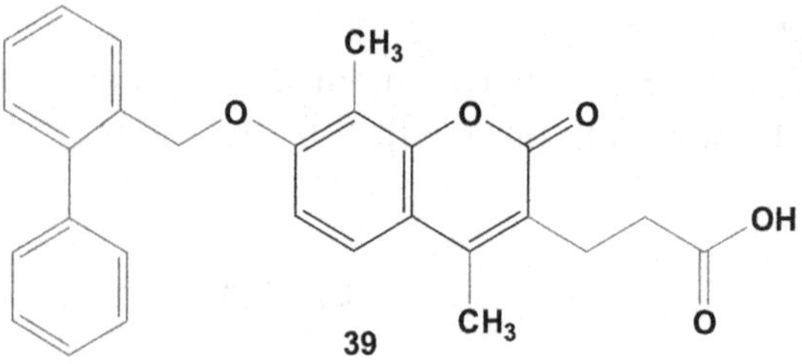

39

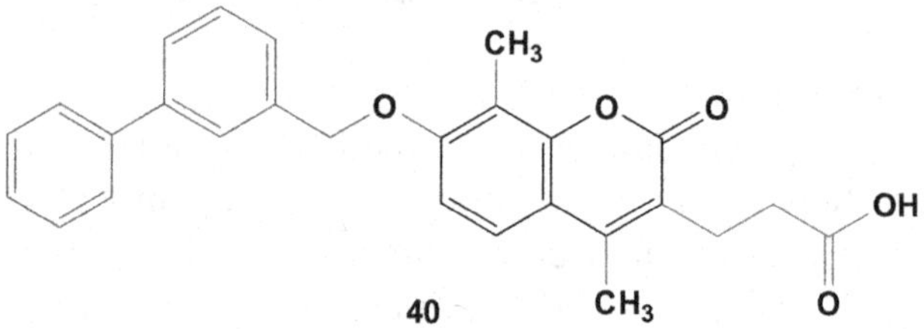

40

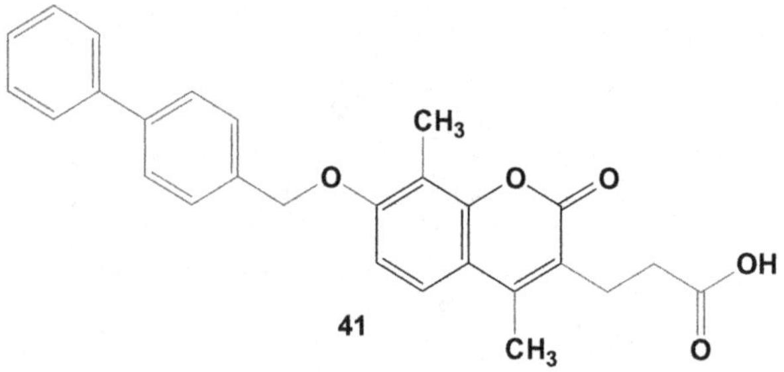

41

Napthyl and Anthracenyl Group

1-Naphthyl [42], 2-naphthyl [43], or anthracenyl [44] substituted compounds were moderately potent, with a range of IC50 values 6 to 10 μM.[17]

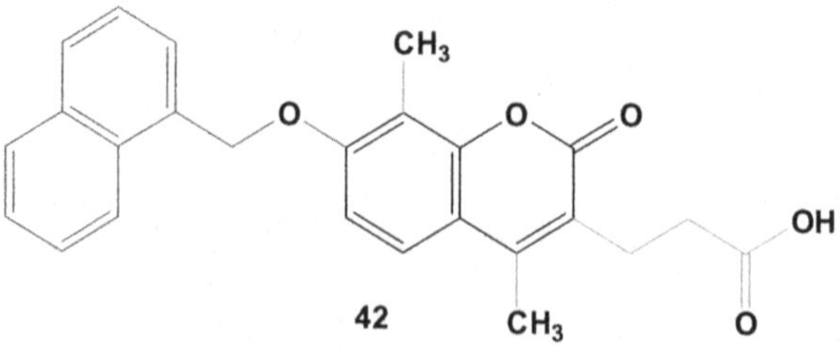

42

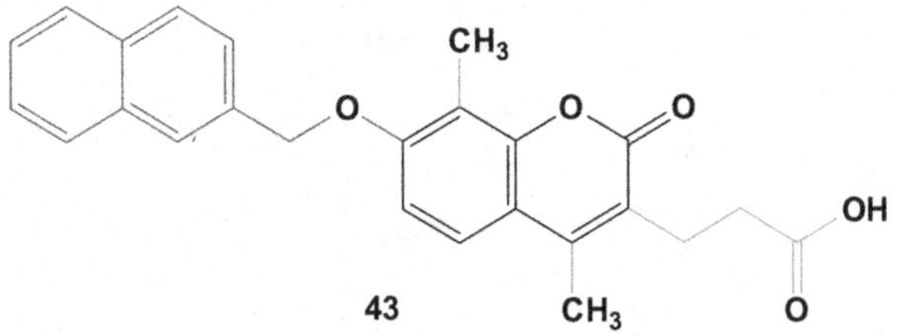

43

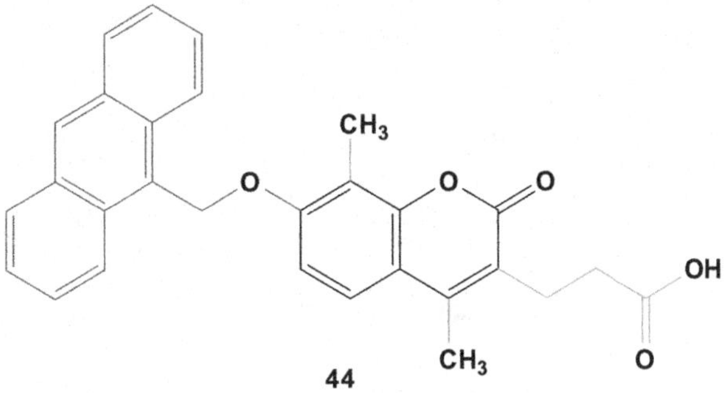

44

Pyridyl Group

No significant helicase inhibitory activity against *S. aureus* and *B. anthracis* was shown by the compound bearing a pyridyl group [45], substituted with an alkoxy derivative at the C-7 position of coumarin ring.[17]

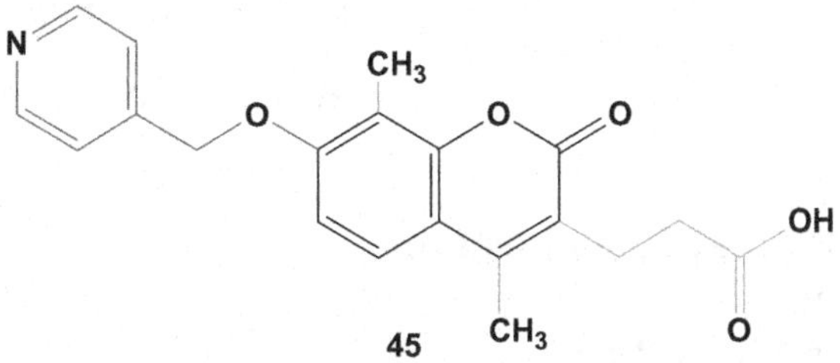

45

Quinolinyl or Isoquinolinyl Group

A weak or no-inhibitory activity for DNA helicases was shown by the compounds that had quinolinyl [46-47] or isoquinolinyl [48-49] substitution.[17]

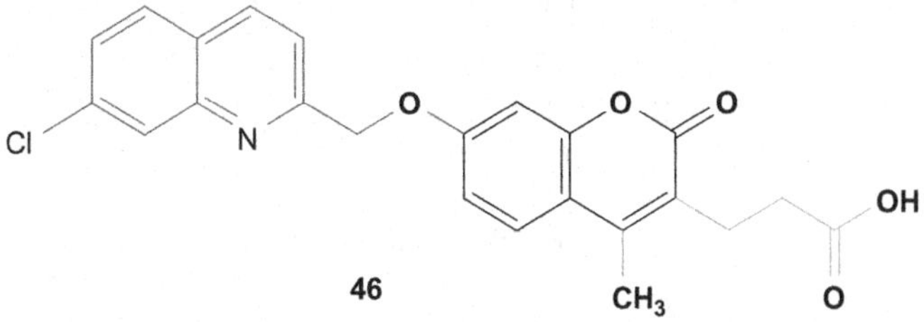

46

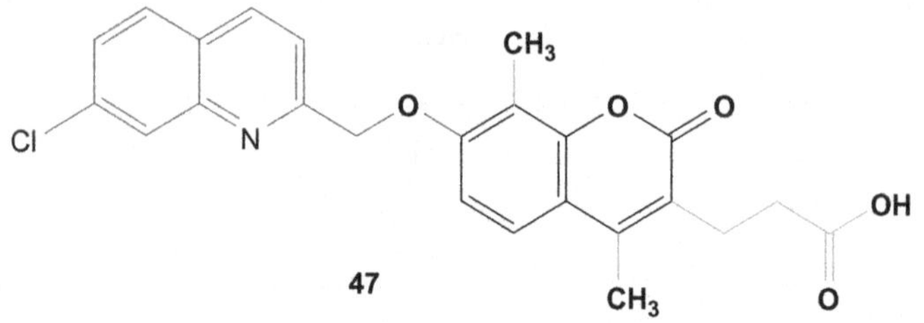

47

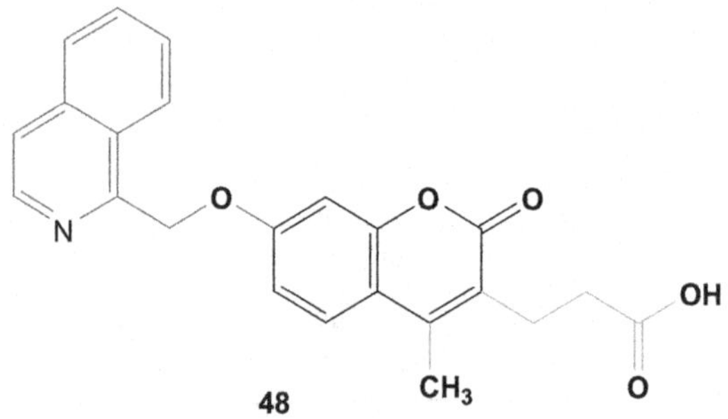

48

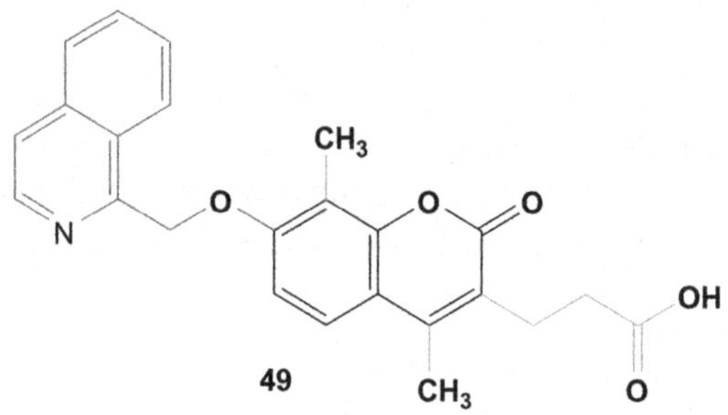

49

2.2.5 Fused Ring at C_6-C_7 Position
2.2.5.1 Furan Ring
Substitution on Furan Ring

The antimicrobial activities of modified psoralen [50] against *H. pylori* were evaluated. The bromine and chlorine substituents of psoralen exhibited inhibitory activity against *H. pylori* at the same level as lead (psoralen) compound, whereas the compound with NO_2 substituent stood two fold stronger than the lead compound.[21]

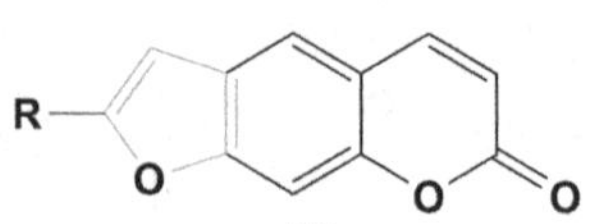

50
R = Br, Cl, NO_2

The modified analogs of psorlen bearing the CH_2R group [51] were inactive against *H.pylori*.[21]

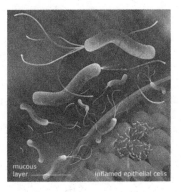

51

R = OMe, OEt, OBu-n

OAc, NEt$_2$

The modified analog of psorlen [52] with the methyl group at C-3 showed decreased activity against *H. pylori* compared with lead.[21]

52

Helicobacter pylori is a gram-negative bacterium found in the stomach. It is present in patients suffering from prolonged gastritis and ulcers. It is well associated with the growth of other intestinal ulcers and abdominal cancer. [25]

Figure 2.5 Helicobacter pylori. www.bio.davidson.edu/people/sosarafova/assets/bio307/vinardone/page01. html.

2.2.6 Functional Group at C_8 Position
2.2.6.1 Dimethyl Group with One Methyl Group at C_7 Position

The outcome of the presence of the methyl group at the C-4 and C-8 positions of the coumarin nucleus was observed by comparing unsubstituted and ethyl-substituted derivatives. Compounds that are unsubstituted at the 8-position of coumarin exhibited two- to three fold decreased potency. This was found by comparing, respectively, [36] vs. [39], [37] vs. [40], and [38] vs. [41] derivatives.[17]

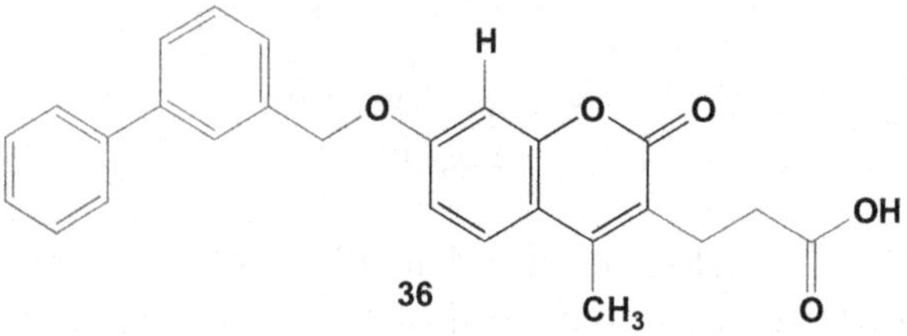

36

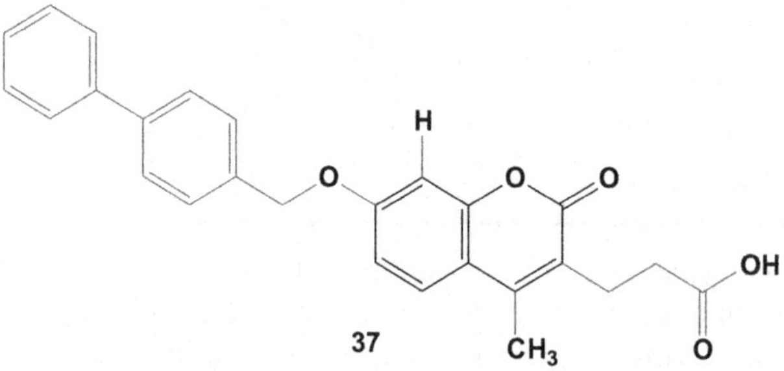

37

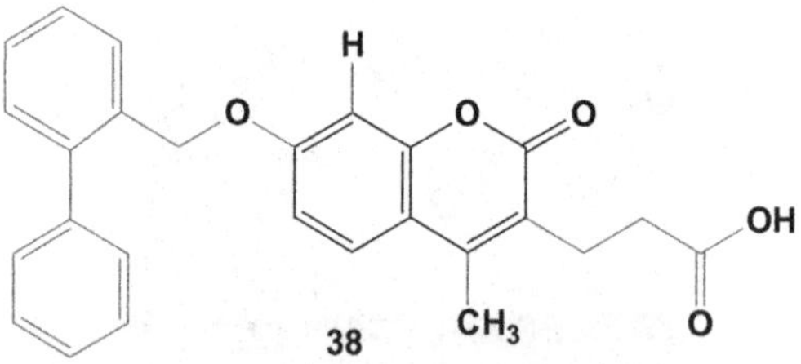

38

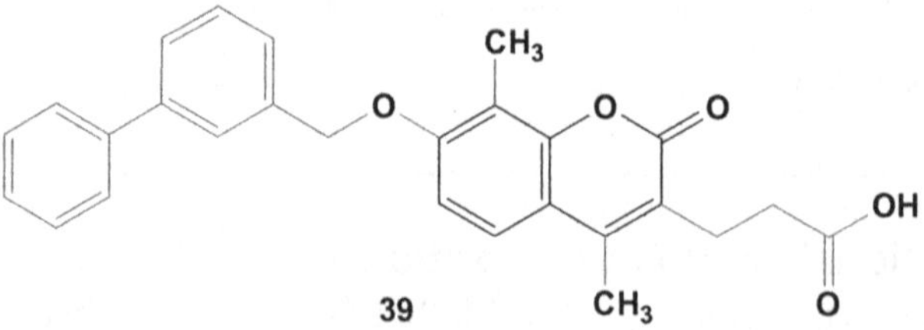

39

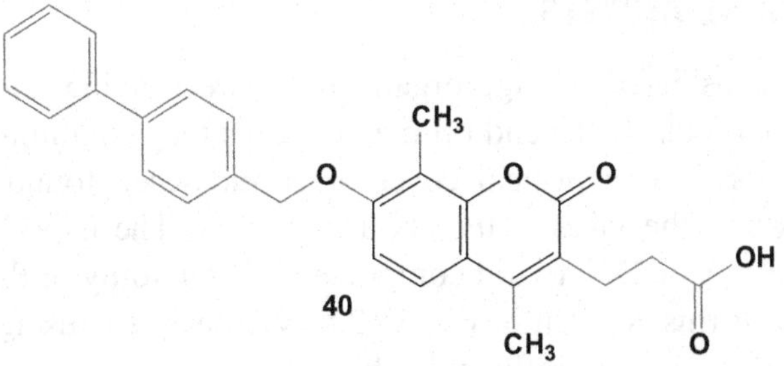

40

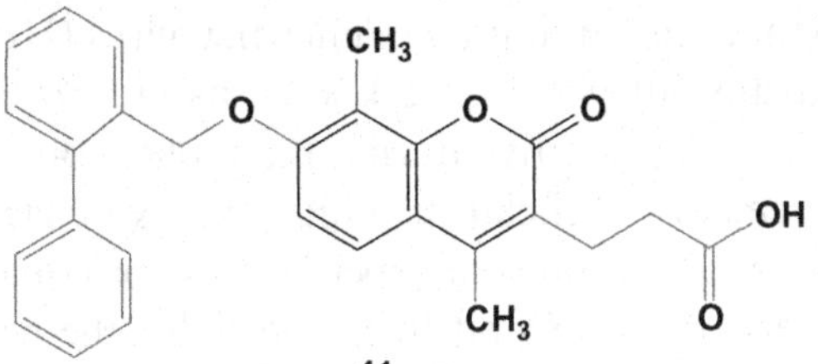

41

2.2.6.2 Ethyl Group

Substitution of the ethyl group at the C-8 of coumarin was well accepted, such as in [53, 54], as evidenced by the low micromolar IC50 values for compounds [35] and [36].[20]

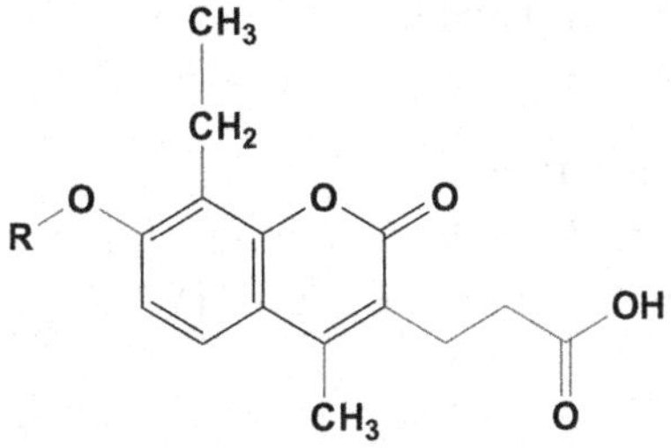

53

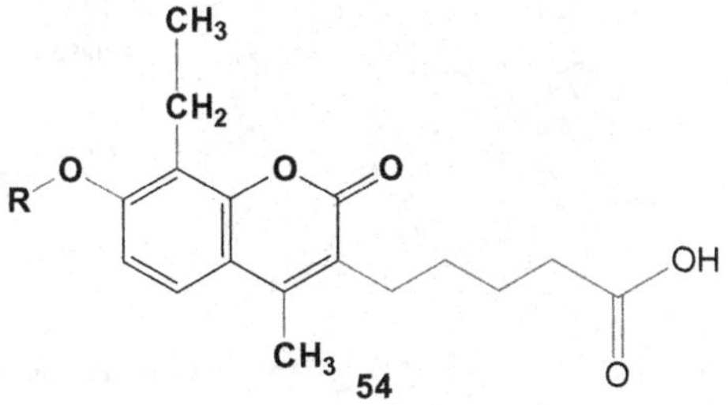

54

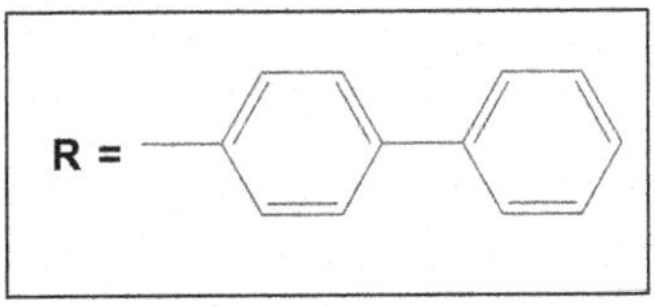

R =

2.3 ANTIVIRAL AGENTS

Viruses are considered linkage organisms between living and nonliving beings and particularly depend on the host cells they contaminate to replicate. They present in protein cover or capsid when found outside of host cells where the virus is metabolically inert. The capsid surrounds with either RNA or DNA viral components.[26] Anatomy of the influenza virus in such forms is given in Figure 2.6. Viruses are causing agents for several diseases, such as influenza, chicken pox, viral pneumonia, viral hepatitis (HCV), and human immunodeficiency virus (HIV). Agents that inhibit the production of viruses are known as *antiviral agents*. Several antiviral agents are simply active during the replication of genetic material. They may remain beneficial in initial phases of viral infections or stop reappearances or else prevent repetition for prolonged infections.[27] It seems that several drugs exhibit their effects for the period of definite phases of viral replication; in addition, many are comparatively toxic for the host also when used for long periods. In wide-ranging remedy of viral infection in animals is unusual with restricted applications, generally in current dealing of infections with ophthalmic. Viruses uses numerous metabolic procedures of the host body, so it is hard to find drugs that are specifically used for viruses. Still, some enzymes found in viruses are effective targets for antiviral agents.

2.3.1 Planarity of Molecules

Planarity in coumarin derivatives is one of the important structural features for development of active anti-HIV agents. The complete loss

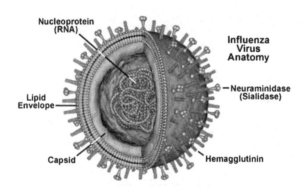

Figure 2.6 Anatomy of virus. www.csa.com/discoveryguides/avian/review2.php.

of anti-HIV activity was observed in those coumarin derivatives in which steric effects occur mostly at C_4 and C_5 substituents. This was expected due to reduction in overall planarity, hence its cause for decrease in resonance of the coumarin ring.[28]

2.3.2 Functional Group on Coumarin Nucleus
2.3.2.1 Small Aliphatic Group
Small aliphatic substitutions such as the methyl group on the coumarin ring [55] are well accepted for enhanced anti-HIV activity, whereas aromatic groups are not effective.[28]

55	R_1	R_2	R_3	R_4
a.	CH_3	CH_3	H	H
b.	H	CH_3	CH_3	H
c.	H	CH_3	H	CH_3
d.	CH_3	H	CH_3	H
e.	Cl	CH_3	H	H
f.	Ph	CH_3	H	H
g.	$-(CH_2)_4-$		H	H
h	H	$CH(CH_3)_2$	CH_3	H
i.	H	CH_3	OCH_3	H
J.	H	CH_3	$OCH_2 C_6H_5$	H

55

Some small substituents other than the alkyl group, such as the OMe [56], Br [57], and OAc [58] groups attached with a coumarin nucleus, produce reduced anti-HCV activity as well as cytotoxicity.[29]

56   OMe

57

58

2.3.3 Functional Group at C_3 Position

2.3.3.1 Benzimidazole-Coumarin with Various Substituents

Benzimidazole–coumarin hybrids, along with their D-ribofuranosides derivatives, were tested for antihepatitis C virus (HCV; see Figure 2.7) activity.[29]

Hepatitis C virus (HCV) is the main causative agent of hepatitis C, a transmittable disease mainly affecting the liver. Chronic infection can lead to damaging of the liver and ultimately to cirrhosis, which is commonly seen after many years of infection. In some cases, those with cirrhosis will go on to develop liver cancer or liver failure.[30]

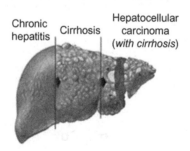

Figure 2.7 HCV infected liver. www.pharmatutor.org/articles/computational-analysis-hcv-entry-inhibitors-hepatitis-c-treatment-molecular-docking-approach

Methyl Group

Benzimidazole-coumarin nucleus substituted with methyl group hybrids such as [59] and [60] exhibit significant anti-HCV activity. The addition of a second methyl group [61] to the benzimidazole nucleus was also expected to enhance the anti-HCV activity of compounds.[29]

61	R₁	R₂	R₃	R₄	R₅	R₆
a.	Me	H	Me	H	H	H
b.	Me	H	H	H	H	OMe
c.	Me	Me	H	H	H	H

Aromatic Group

The addition of a fused benzene ring with a coumarin nucleus [64] improves the anti-HCV activity, whereas its addition with a benzimidazole nucleus (substituted with a coumarin ring [62-63]) leads to inactive or very little active anti-HCV compound.[29]

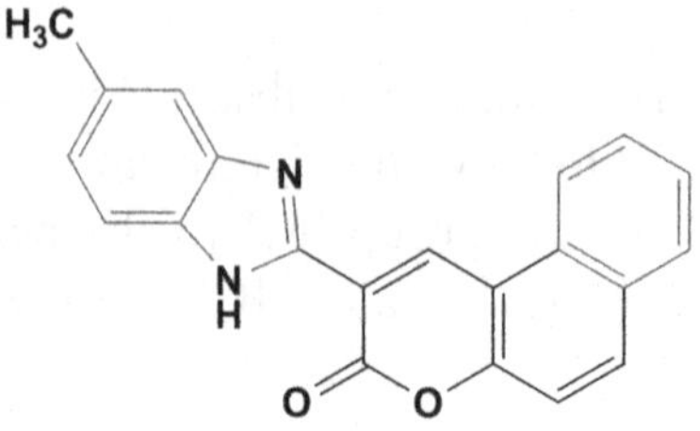

64

D-ribofuranose Group

Highly conjugated compounds were obtained through substitution of D-ribofuranose moiety onto the benzimidazole-coumarin derivative [65]. This highly conjugated system was found to be effective as an anti-HCV compound.[29]

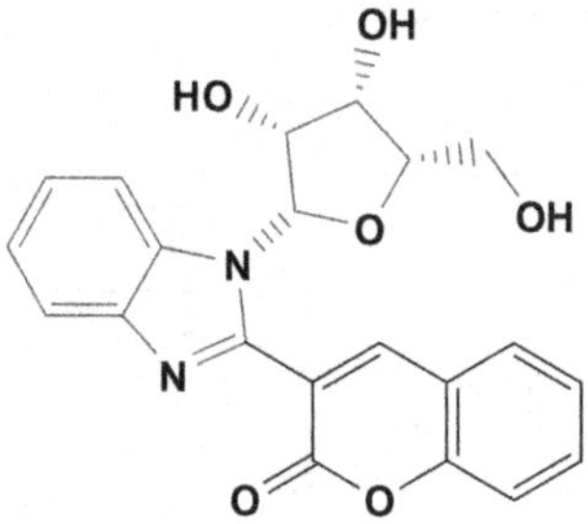

65

D-ribofuranose moiety with the absence of acetyl groups [66, 67] enhances the proficiency for anti-HCV activity.[29]

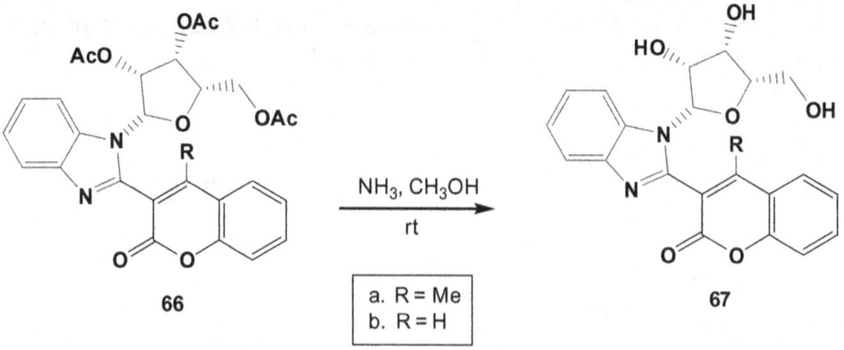

NH_3, CH_3OH

rt

| a. R = Me |
| b. R = H |

66   **67**

2.3.3.2 SCH$_2$ Linkage

Benzoxazole and Benzothiazole Group

Attachment of benzoxazole and benzothiazole nuclei [68a-e] with the SCH$_2$ terminal of the coumarin nucleus was potent to exhibit anti-HCV activity, having EC50 value of 6.8-12 μM and significant selective features. When benzoxazole moiety replaces the Br-group on the coumarin nucleus, it causes a decrease in cytotoxic activity by 2.9-fold.[31]

68	X	R_1	R_2
a.	O	H	H
b.	O	H	OMe
c.	O	Br	H
d.	S	H	H
e.	S	H	OMe

Imidazopyridine Group

The presence of the imidazopyridine nucleus [69] at the SCH$_2$ terminal of the coumarin moiety improves HCV inhibition by a factor of 7.4. The S-CH$_2$ linkage group is essential and must not be removed. Substitution of the methoxy group on the coumarin ring with the imidazopyridine moiety developed antiviral activity 5.4-fold.[31]

Substitution of bromine on the coumarin ring with the imidazopyridine moiety [70] enhanced the HCV inhibition by 8.7-fold.[31]

A carbon atom of the imidazopyridine ring replaced with a nitrogen atom gives purine. An increase in HCV inhibitory activity by 29-fold was found in coumarin substituted with purine [71a-b].[31]

71	X	R_1	R_2
a.	CH	H	H
d.	N	H	H

Benzimidazole Group

Coumarin moiety with a benzimidazole nucleus [72] that has an -S-CH$_2$ linker is the important factor in increasing anti-HCV activity. After comparing compounds [72, 73, 74] it was found that addition of the -S-CH$_2$ linker between the coumarin moiety and the benzimidazole nucleus enhanced the HCV inhibitory activity by a factor of 12.[32]

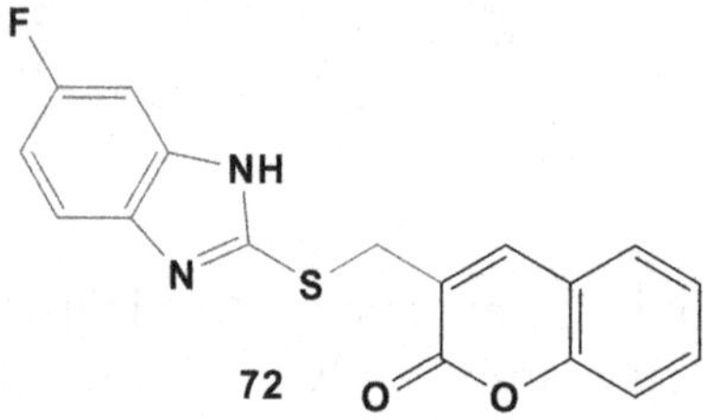

72

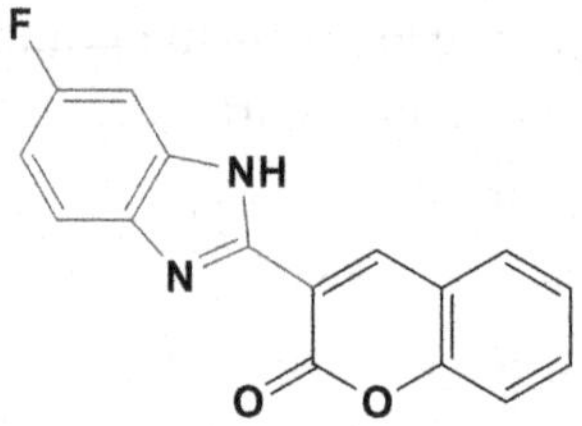

73

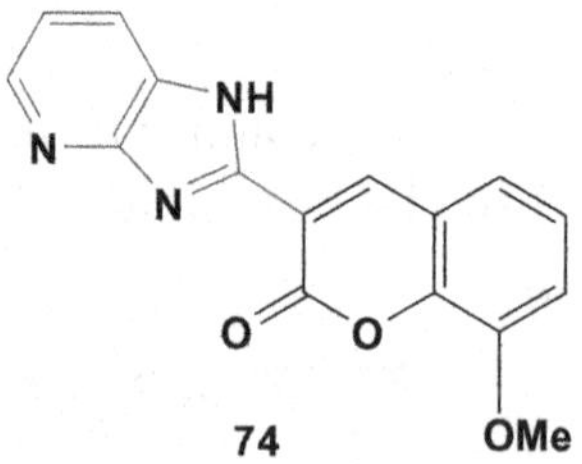

74

The presence of the peracetyl 2-deoxy-D-glucose group on the benzimidazole-coumarin ring improved cytotoxicity and HCV inhibition by 8.2-fold. This was proven by comparison of compounds [75] and [76].[31]

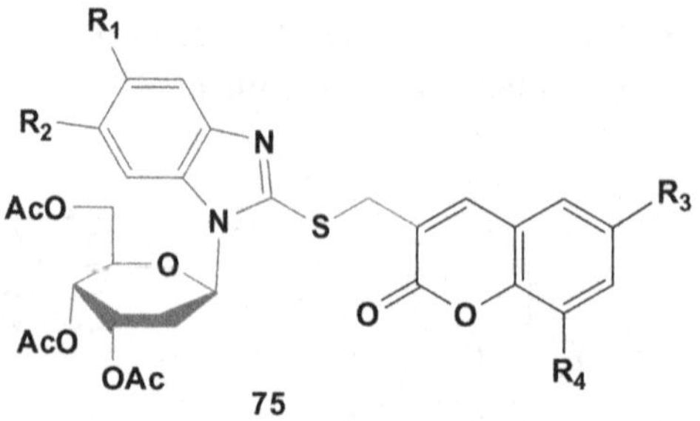

75

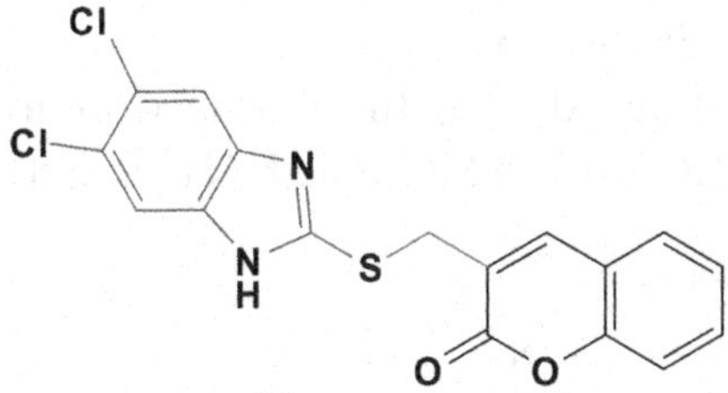

76

Guanosine Group

Replacement of the benzimidazole moiety [72] with a guanosine moiety [77] in conjugated coumarin derivatives has no significant inhibitory activity on HCV.[32]

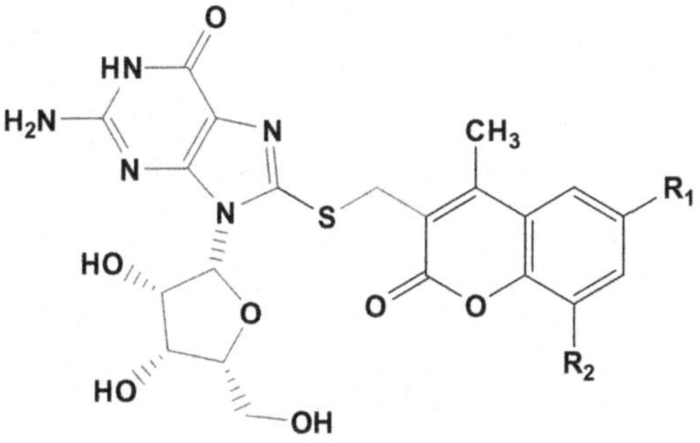

77

Adenosine Group

Addition of an amino group on purine moiety gave the coumarin-adenosine derivatives [78a-d] but caused the reduction of inhibitory activity for HCV as well as SI values. This was revealed by comparing the analogous coumarin-purine derivatives.[33]

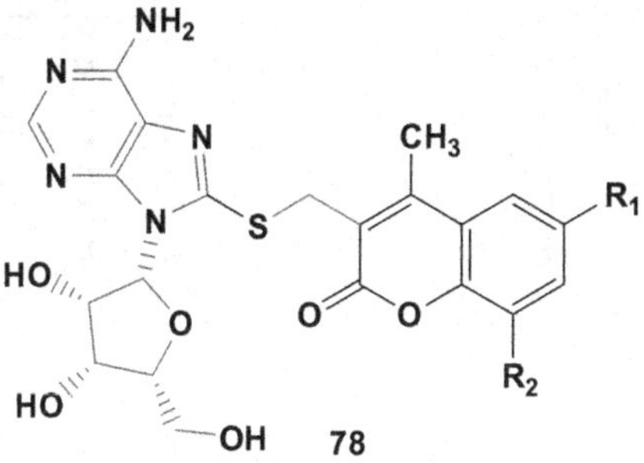

78

78	R₁	R₂
a.	H	H
b.	Br	H
c.	Br	Br
d.	H	OMe

Purine Ribofuranoside Group

Coumarin moiety attached with the purine ribofuranoside [79, 80] is the important structure to increase anti-HCV activity.[33]

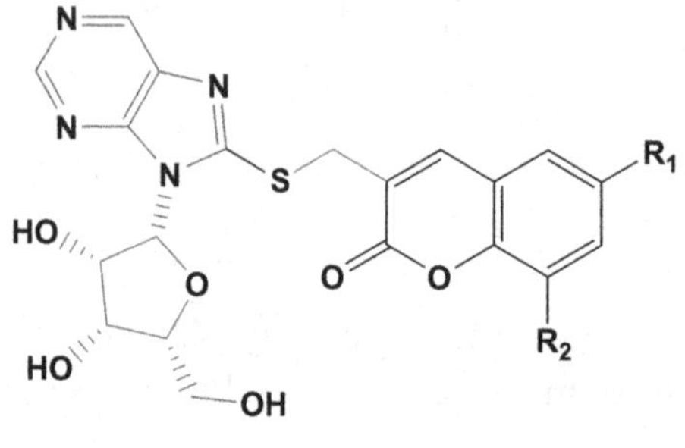

79

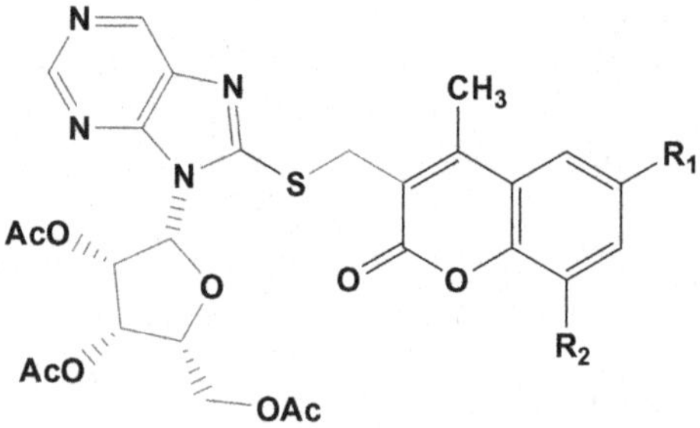

80

No change or very small changes were found in the anti-HCV potency of various substituents of the electron-withdrawing group as well as electron-releasing groups such as Me, F, Cl, Br, and OMe on coumarin-ribofurano derivatives [79a-g] with EC50 values of 4.3−9.5 μM.[33]

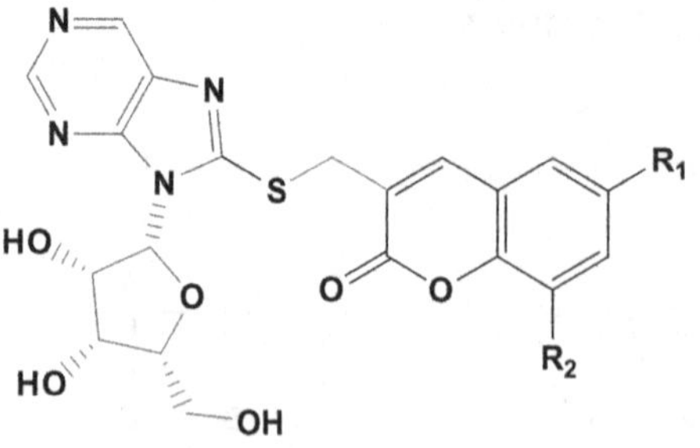

79

79	R_1	R_2
a.	H	H
b.	Me	H
c.	H	OMe
d.	F	H
e.	Cl	H
f.	Br	H
g.	Br	Br

Purine Ribofuranoside with Free −OH Group

In purine-ribofuran derivatives, the presence of free -OH groups at 2'-, 3'-, and 5'-positions, as shown in compound **[81]**, caused reduction of cytotoxicity, whereas the SI values were increased.[33]

2.4 ANTIFUNGAL AGENTS

Fungi, shown in Figure 2.8, are the most common infective agents and can cause several diseases. *Antifungal agents* are the drugs that eradicate fungal infections from host bodies through least-toxic effects. Antibacterial drugs have followed the improvement of antifungal drugs due to the complex basic structure of the fungal organism. Bacterial cells are of a prokaryotic (no karyons) nature. Therefore, their several physical as well as

Figure 2.8 Image of Fungi. http://peapod-peasinapod.blogspot.in/2013_09_01_archive.html

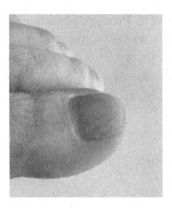

Figure 2.9 Fungal infection of Toenail (distal subungual onychomycosis). http://www.botavie.com/nail-mycosis-s-59.html

metabolic activities are entirely different from those of other human hosts. In contrast, fungi are eukaryotes; thus several drugs lethal to fungi remain lethal for the host body as well. Moreover, fungi usually develop gradually in multicellular systems, so they are difficult to estimate compared to bacteria. This makes it difficult to investigate designing and estimating the *in vivo* or *in vitro* belongings of effective antifungal agents. In spite of these restrictions, several developments have been completed in emerging new antifungal drugs and in understanding the remaining derivatives. In antifungal medications preventing ringworm, athlete's foot[34], and onycho-mycosis[35] (see Figure 2.9), antifungal agents derived from coumarin were commonly synthesized. Herein information regarding the SAR of antifungal coumarins is illustrated.

2.4.1 Functional Group at C_3 Position
2.4.1.1 Nitro Group
Nitration of the inactive coumarin compounds [82] resulted in one of the most active derivatives among the series.[36]

82

2.4.2 Functional Group at C$_4$ Position
2.4.2.1 Acetyl Group

4-hydroxycoumarin derivatives that brought about acetylation [83] are the most active compounds from each series.[36]

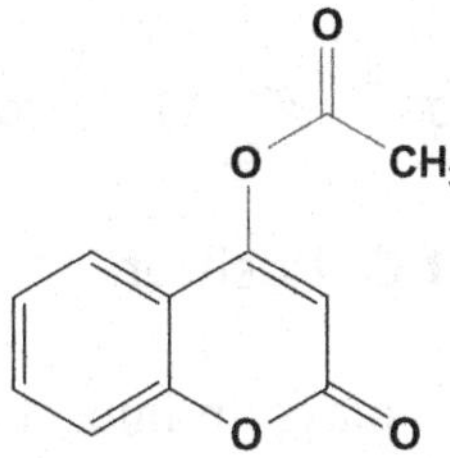

83

2.4.3 Functional Group at C$_6$ Position
2.4.3.1 Ether Group

Substitutions at the C-6 position of coumarin [84-86] did not show any increase in fungicidal activity. Thus all derivatives are inactive.[36]

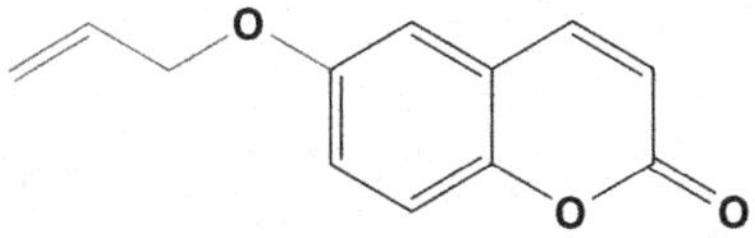

84

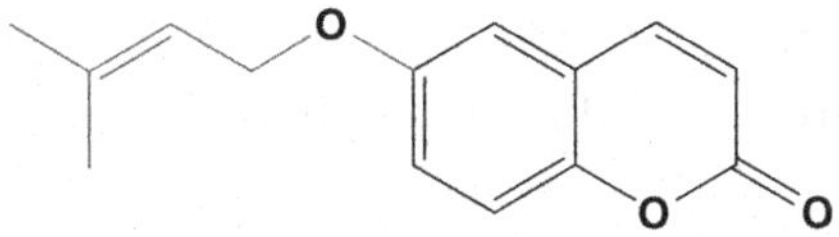

85

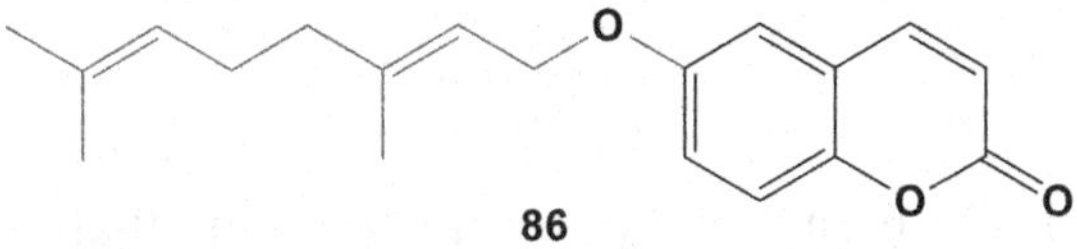

86

2.4.3.2 Acetyl Group

Acetylation of 6-hydroxycoumarins [87] were found to be the most active compounds in each respective series.[36]

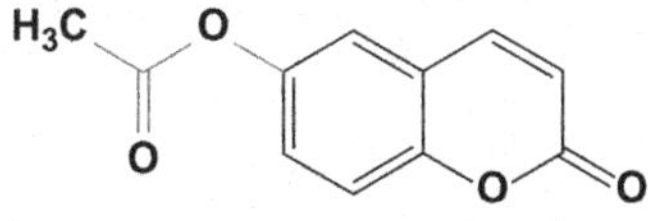

87

2.4.3.3 Nitro Group

Nitration of the compounds [88] resulted in the active antifungal compound.[36]

88

2.4.4 Functional Group at C_7 Position
2.4.4.1 Hydroxy Group

Size of the substituted group plays an important role in the 7-hydroxy-coumarin series to increase fungicidal activity. Thus, decrease in length of the alkenyl side chain increases the fungicidal activity.[36]

Small Aliphatic Chain
Compounds having the smallest aliphatic chain, such as compound [89], showed the greatest activity in the series.[36]

89

Prenyl Group
Compound [90] using an intermediate side chain (especially the prenyl group) displayed strong to adequate activity.[36]

90

Large Chain
The long chain of a prenyl group, as in compound [91], is found absolutely inactive for fungicidal activity.[36]

91

2.4.4.2 Acetyl Group

Compounds resulting from acetylation of 7-hydroxycoumarins [92] are the most active compound among the resultant set of derivatives.[36]

92

2.4.4.3 Nitro Group

Nitration at C_7 of the coumarin of inactive compounds resulted in the corresponding most active compound [93].[36]

93

2.4.4.4 Thiosemicarbazide and Thiazolidinone Group

Coumarin derivatives 4-thiazolidinones and thiosemicarbazides, both with alkyl groups, do not display any important antifungal activity on *A. flavus*. However, on comparing the two derivatives 4-thiazolidinones and thiosemicarbazides, 4-thiazolidinones exhibited improved antifungal activity compared with 4-thiosemicarbazides toward *A. ochraceus*, which highlights the importance of 4-thiazolidinone as well as phenyl rings for achieving antifungal activity. Hence, compound [94-97] exhibit prominent antifungal activity.[37]

94

95

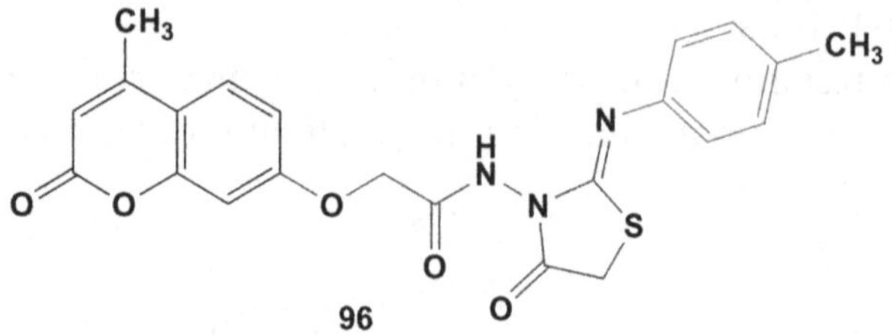

96

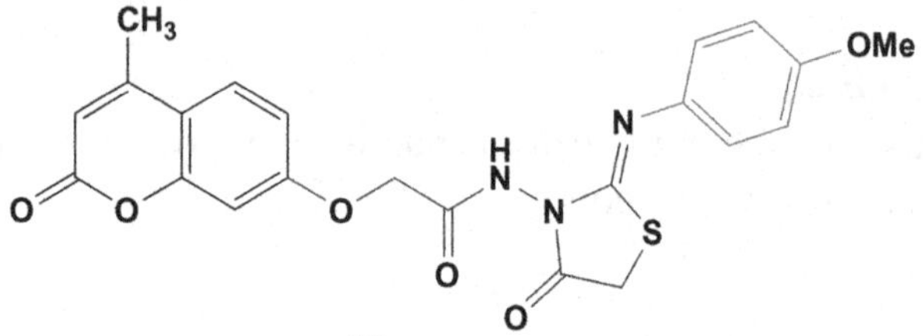

97

It was also revealed that the compounds substituted at the 7-position of 4-thiazolidinone and thiosemicarbazide moieties developed antifungal activity, especially against *F. verticillioides*.[37]

Phenyl Group

The presence of a phenyl ring in compounds [98-99] was found to be a significant antifungal agent against *A. ochraceus*. Unsubstituted or alkyl-substituted phenyl rings did not express any significant increase in antifungal activity toward *F. verticillioides*. But such compounds displayed an admirable antifungal activity against *F. graminearum*. Hence, *F. verticillioides* species was the minimum at risk toward the tested compounds, and the compounds substituted with phenyl rings are the best antifungals.[37]

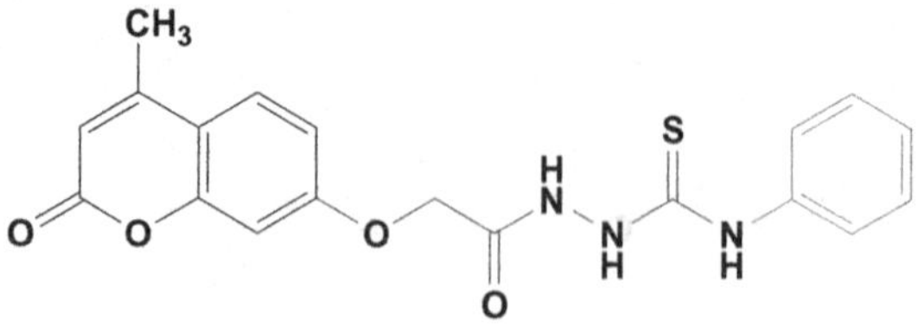

98

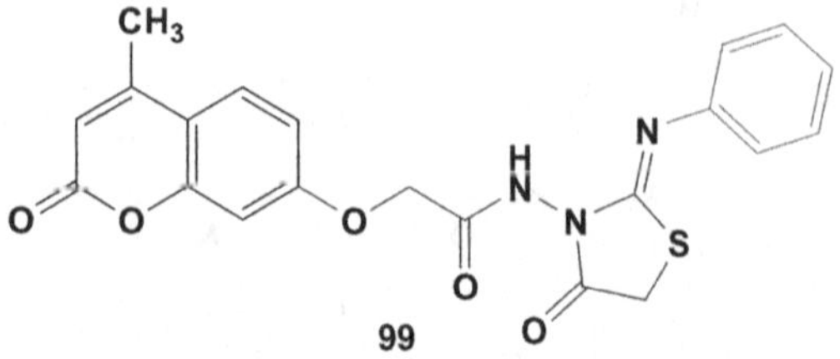

99

Commonly occurring fungi *Fusarium* (species) produces fumonisins. Fumonisins are hepatotoxic (affecting the liver) to entire animal species and have a number of additional effects that reduce the performance of husbandry animals. In harvest, the crop most affected by fumonisins is corn and barley.[38]

Figure 2.10 Fusarium (Species). www.wattagnet.com/Are_fumonisins_really_a_problem_in_poultry_ feed_.html.

CHAPTER 3

Antitumor Agents

Santhosh Penta

Department of Chemistry, National Institute of Technology, Raipur, Chhattisgarh, India

3.1 INTRODUCTION

Cells are the basic units of life. All cancers initiate within cells, so it is necessary to know the changes in normal cells that indicate they are becoming cancerous cells. Normal cells develop and divide in an organized approach to generate more cells to nourish the body. When these cells are damaged or old, they are either replaced by new cells or they die. Still, sometimes this systematic development goes wrong. The DNA (the genetic material) of a cell can become mutated, which affects the normal growth of the cell and its division. When cells do not die in obligatory conditions and when new cells forms without the need of a body, the additional cells may form a group of tissues called a *tumor*. A tumor, or mass of cells, begins to proliferate in its own way. The whole process is well explained in Figure 3.1.

Antitumor agents prevent or inhibit the formation or growth of tumors and are known as antitumor, anticancer, chemotherapeutic, or antimetastatic agents. Antitumor agents kill those cells that divide rapidly, which is one of the main property in most types of cancer cells. This phenomenon indicates that chemotherapy is harmful for all those cells that divide rapidly in normal conditions—for example, cells in the digestive tract, bone marrow, and hair follicles. These outcomes are the most common problem of chemotherapeutic agents. Cytotoxicity of anticancer agents should be in a considerable range.[39]

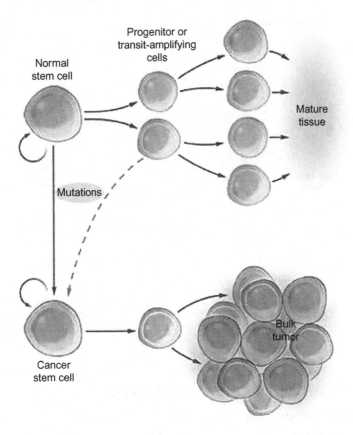

Figure 3.1 Cell Division of Normal & Cancer cell. www.jacquieforall.com/Understanding%20ALL.php.

3.2 UNSUBSTITUTED COUMARIN

The unsubstituted coumarin showed medium to weak hCA II inhibition.[40] This is the first example in the collected work that does not inhibit hCA II. This isoform is specified particularly to sulfamate and sulfonamide inhibitors, which pointed to drugs having numbers of adverse effects.[41]

In the coumarin nucleus, substituents at position C-3, C-4, or C-7 enhance biological activities, and they are recognized to bring apoptosis in leukemic cells through activating the cysteine protease 32 kDa proenzyme and enhancing cytochrome C. It is also proposed that the proper substitution at the 3 and/or 4 position of the coumarin molecule is

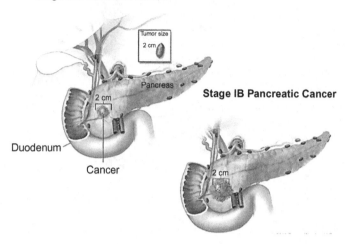

Figure 3.2 Pancreatic Cancer Cells. www.cancer.gov/cancertopics/pdq/treatment/pancreatic/Patient/page2.

essential for designing effective cytotoxic agents. The electron-donating and -accepting effects are significant factors for coumarin derivatives in estimating the cytotoxic action.[40]

3.3 SUBSTITUTION IN COUMARIN NUCLEUS

3.3.1 Hydroxycoumarin

Hydroxycoumarin derivatives were expected to have activity in contradiction of human pancreatic cancer cells (PANC-1) as in Figure 3.2 in nutrient-deprived conditions. The simple coumarin moeity [100-101a] showed no appreciable cytotoxicity, and compounds [101b-c] also did not display any cytotoxicity when shown the existence of an iodo group or benzyl group is insignificant for the biological activity.[42]

101	R_1	R_2	R_3
a	OH	I	H
b	$-O\!-\!C_6H_5$	H	H
c	$-O\!-\!C_6H_5$	I	H

100

101

Compound **[102]**, having a small acyclic 4-carbon alkyl ether chain, does not exhibit any considerable activity.[42]

102

Compound **[103]**, with a cyclic alkyl chain of 5-carbon, showed good cytotoxic activity against PANC-1 cells in nutrient-deprived conditions at a concentration of 200 1M.[42]

103

Compound **[90]** showed good cytotoxic activity against PANC-1 at a concentration of 100 1M. Compounds **[103]** and **[90]** both retain a 5-carbon ether chain. It is expected that experiential cytotoxic activity is due to an increase in hydrophobicity by means of added carbon on the ether chain and the isoprenyl moiety.[42]

90

Isoprenyloxy derivatives **[91,104-105]** suggested that the development of cytotoxic activity is related to the existence of longer ether chains. The specific position of the isoprenyloxy chain is not essential for cytotoxic activity.[42]

91

104

105

3.3.2 Allyl/Ether Group

Allyl-substituted coumarins **[106a-c]** were effective hCA IX inhibitors. The position of allyl/ether groups intensely affects the inhibitory activity; hence, the 4-substituted derivative showed 4.4 times enhanced inhibitor activity related to the 6-substituted compound. The 7-substituted compound showed moderate activities compared with the two resultant isomers.[43]

The carbonic anhydrases (or carbonate dehydratases) constitute a group of enzymes. These are applicable as catalysts for rapid interconversion of carbon dioxide and water to bicarbonate and protons, and vice versa. In the absence of catalysts' reversible reaction, that occurs relatively slowly.

Figure 3.3 Ribben Diagram of Carbonic Anhydrase. http://aci.anorg.chemie.tu-muenchen.de/anwander/lehre_bio_e.html.

| 106a. 6-substituted |
| 106b. 7-substituted |
| 106c. 4-substituted |

106

Coumarin-based natural product **[107]** showed an hCA II inhibitor with a KI of 59 nM.[44]

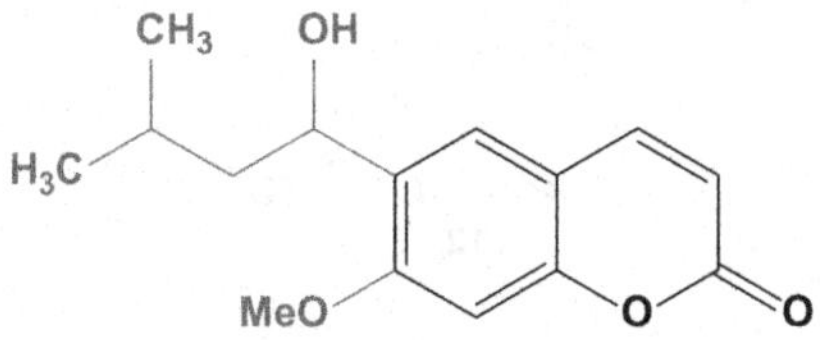

107

3.3.3 Aminocoumarins

Aminocoumarins **[108]** were somewhat effective as hCA I inhibitors and were a weak hCA XII inhibitor compared with other derivatives such as **[109-111]**.[43]

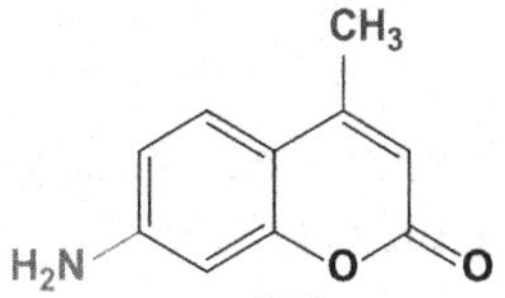

108

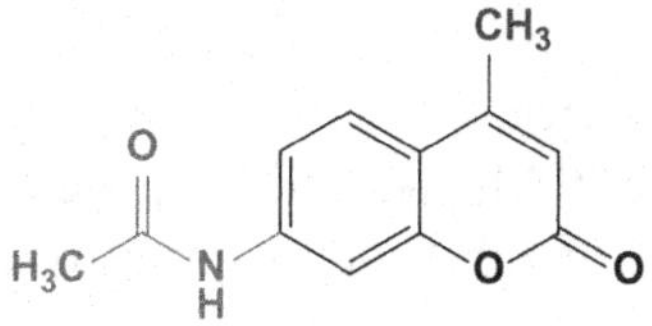

109

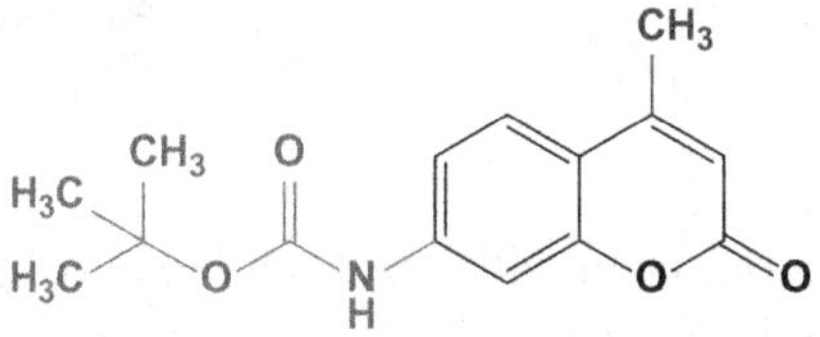

110

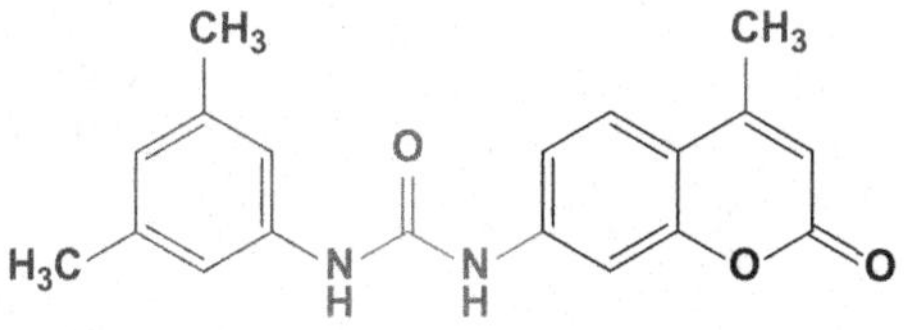

111

3.3.4 Functional Group at C$_3$ Position

3.3.4.1 Hydroxyl Group

Cytotoxicity against the selected cancer cells was confirmed by substitutions of hydroxyl groups in the coumarin molecule at C-3 position **[112]**.[40]

112

3.3.4.2 Amine Group

The influence of complexes of the general formula PdL_2 substituted with coumarin at the N atom was examined on two human leukemia cell lines, NALM-6 and HL-60. Pd complex cis-[**113a**] exhibited important cytotoxic activity. It was also shown that compounds were more toxic for NALM-6 rather than HL-60 leukemia cell lines.[45]

Size of Group
For the NALM-6 cell line, the cytotoxic effect improved by increasing the size of substituents on nitrogen atoms of ligands [**113a-c**].[45]

113

113	R	R₁
a.	Ph	H
b.	Ph	CH₃
c.	Ph	CH₂Ph

The C-3 atom of coumarin with an ethylidene group [**114-115**] showed effective cytotoxic activity with an increase in the size of substituents.[45]

114   **115**

Palladium Complex
The palladium-cis complex, particularly compound [**116**], displayed the maximum cytotoxicity activity among the all palladium toward the HL-60 and NALM-6 cells.[45]

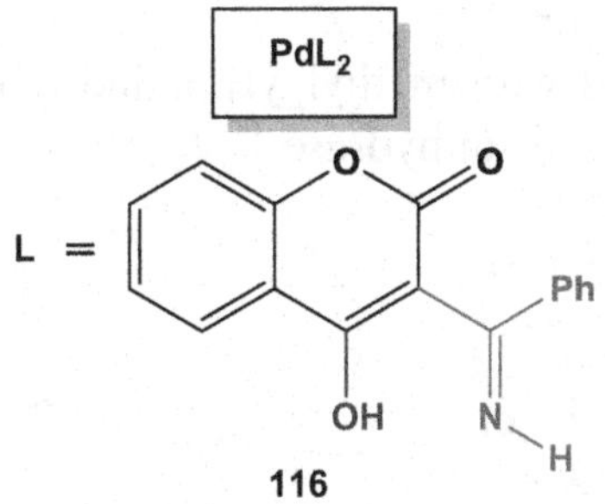

116

Palladium complexes **[117a-d]** have slightly lower cytotoxic effects than carboplatin.[45]

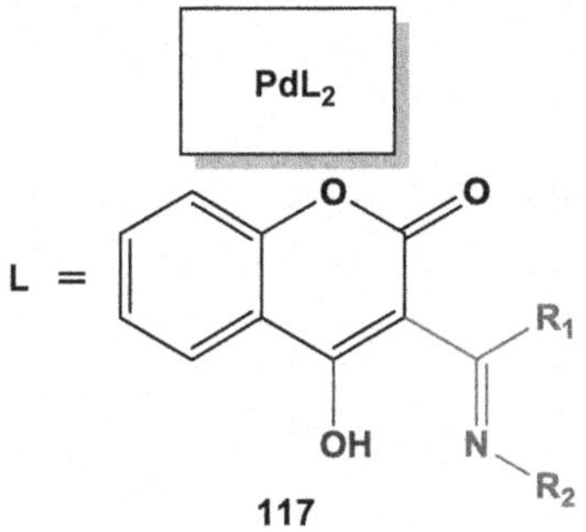

117	R$_1$	R$_2$
a.	Ph	CH$_3$
b.	Ph	CH$_2$Ph
c.	CH$_3$	CH$_3$
d.	CH$_3$	CH$_2$Ph

117

3.3.4.3 Sulphonamide Group

A set of sulphonamides containing coumarin derivatives exhibited a very remarkable profile for two human carbonic anhydrase inhibitors. Sulphonamide compounds without substitution on a benzene ring of the coumarin moiety and the substituted group on the amino derivatives **[118]** were more effective for hCA II, with the effective substitution in order of -H > -CH$_3$ > -tBu. On the other hand, for hCA IX the order is CH$_3$ > -tBu > -H.[46]

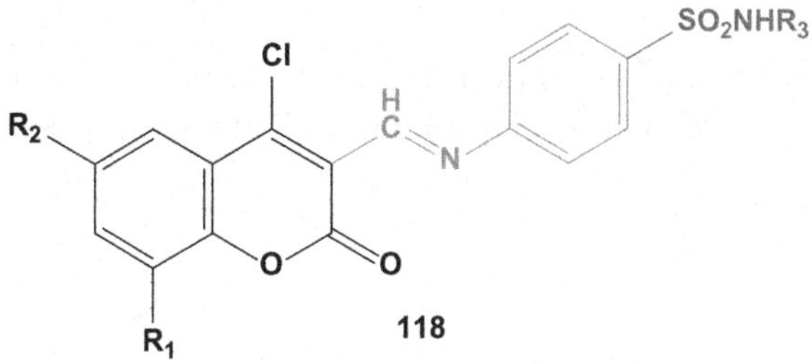

118

Methylpyrimidine Group

Coumarins substituted with methylpyrimidine moieties [119] were the best inhibitors of carbonic anhydrase.[46]

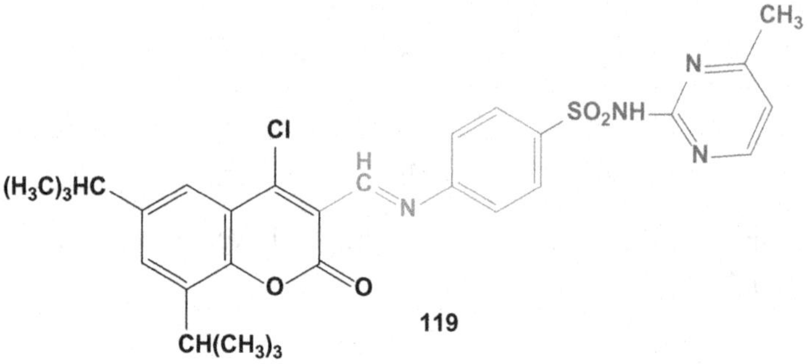

119

Acetimidamide Group

Compound [120] is an acetimidamide derivative and was found to be the least active for hCA.[46]

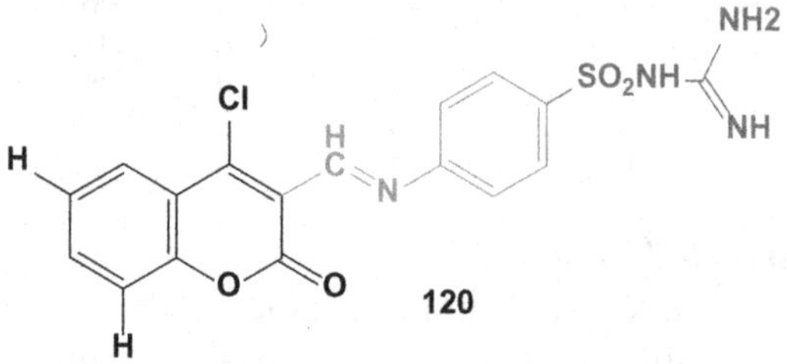

120

3.3.4.4 Hydrazide-Hydrazone Group

Hydrazide-hydrazone-bearing coumarin moiety were evaluated against *in vitro* drug-resistant pancreatic carcinoma cells (PANC-1), leukemia cell lines (CCRF), and drug-sensitive hepatic carcinoma (Hep-G2).[47]

Pyrrole Group

The pyrrole-substituted coumarin hydrazide-hydrazone derivatives [121] were the most active against Hep-G2 and CCRF cancer cell lines.[47]

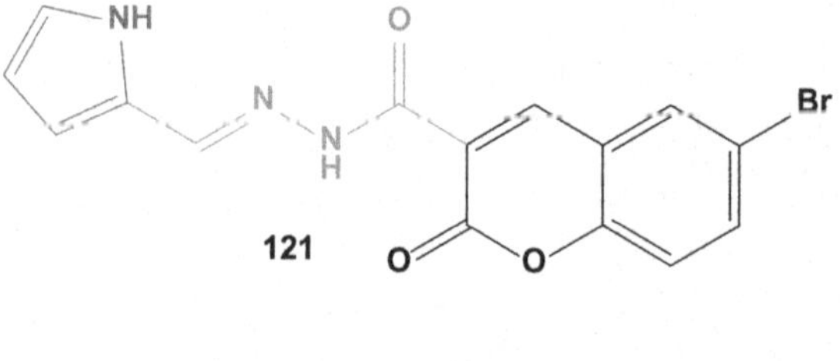

121

Furan Group

Furan and Isatin substituted coumarin hydrazide-hydrazone derivatives [122] were the most active against PANC-1 cell lines, whereas pyrrole derivatives were the least active against the same cell lines.[47]

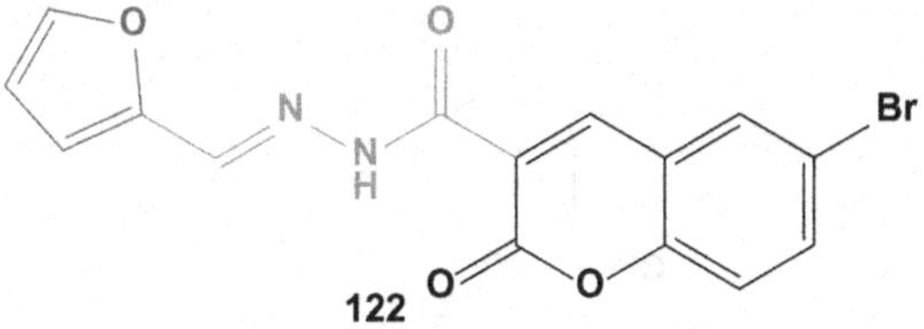

122

Thiophene Group

The thiophene-substituted coumarin hydrazide-hydrazone derivatives [123] were the most active against Hep-G2 and CCRF cancer cell lines.[47]

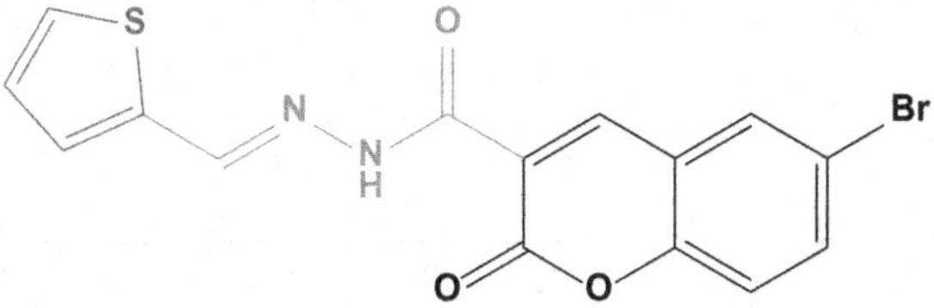

123

Isatin Group [124].[47]

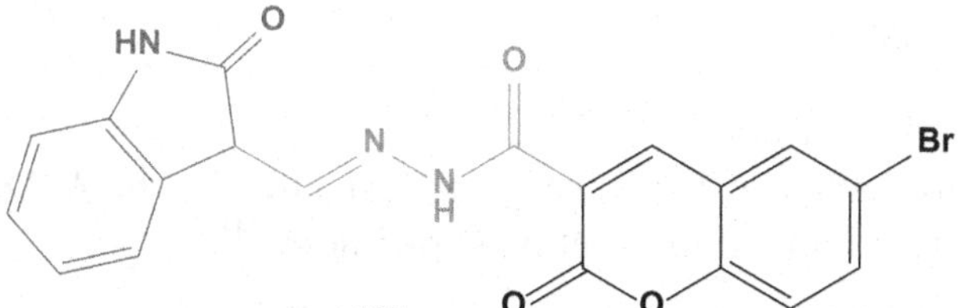

124

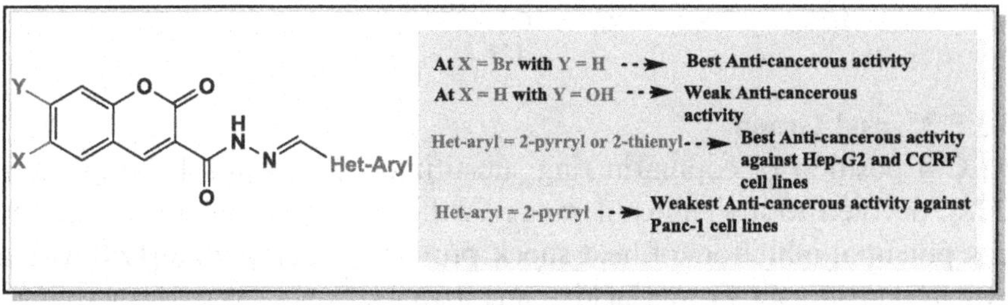

At X = Br with Y = H	Best Anti-cancerous activity
At X = H with Y = OH	Weak Anti-cancerous activity
Het-aryl = 2-pyrryl or 2-thienyl	Best Anti-cancerous activity against Hep-G2 and CCRF cell lines
Het-aryl = 2-pyrryl	Weakest Anti-cancerous activity against Panc-1 cell lines

Figure 3.4 SAR of Anti-cancerous agent.

3.3.5 Functional Group at C_4 Position
3.3.5.1 Hydroxyl Group

OH moiety in the 4-, 6-, or 7-position of the heterocyclic ring, as in compound [125], showed the best activity for hCA IX and hCA XII inhibition.[41]

125

The hydroxyl substitutions in the coumarin molecule at the C-4 position [126a-c] have demonstrated considerable cytotoxicity against the selected cancer cells, such as HL 60 cancer cells and Hep-3B cancer cells lines.[40]

126	R_1	R_2	R_3	R_4	R_5	R_6
a.	H	OH	H	CH_3	H	H
b.	H	OH	H	CH_3	CH_3	H
c.	C_6H_5	OH	OH	H	OH	H

126

3.3.5.2 Chlorine Group

The 4-chlorocoumarin [127] was the least effective hCA IX inhibitor among hydroxyl and chloromethyl substituents.[41]

127

3.3.5.3 Tosyl Group

At C-4 position of coumarin ring substitution with a tosyl group, as in [128], devoted to a significant range for highest antiproliferative activity and potential inhibitors of heat shock protein 90; compound [129] with a hydroxyl group at C-4 was lacking any activity.[48]

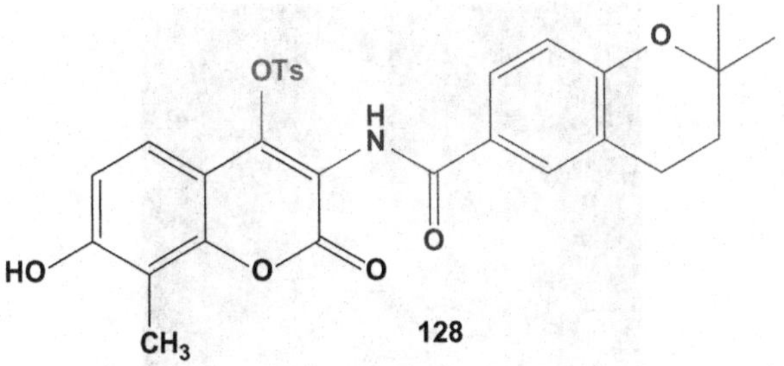

128

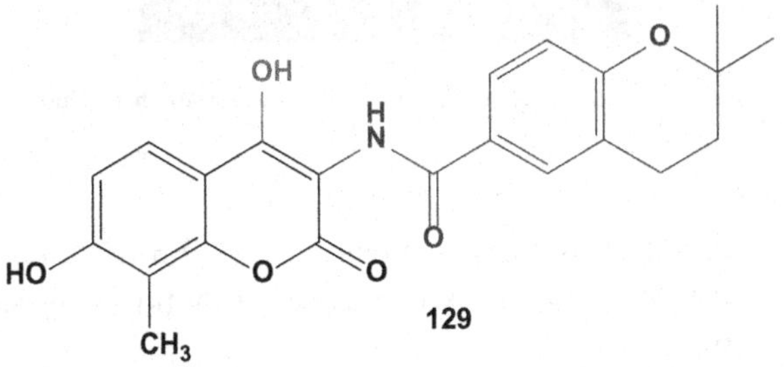

129

3.3.5.4 Triazole Group

4-(1,2,3-triazol-1-yl)coumarin moieties [130] were estimated for *in vitro* anticancer activities against three human cancer cell lines: human breast carcinoma MCF-7 cells, colon carcinoma SW480 cells, and lung carcinoma A549 cells (Figure 3.5).[49]

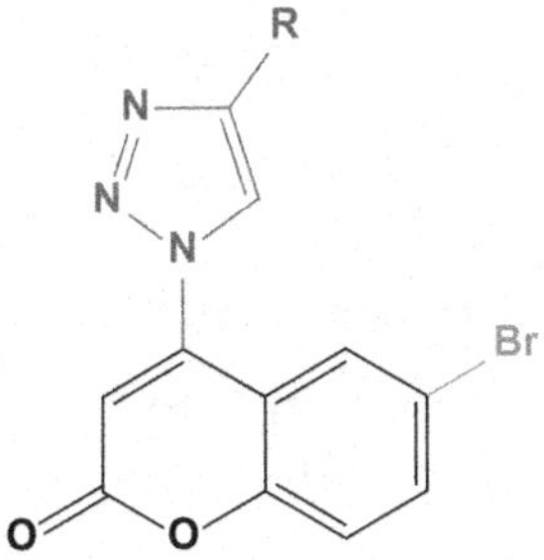

130

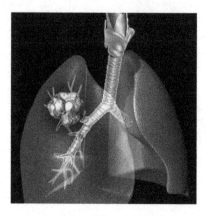

Alkoxy Bridge

The SAR revealed that at the C-4 position of 1,2,3-triazole, $-CH_2-O-$ bridge is the best for bioactivity, hence compound [131] showed good cytotoxic activity.[49]

Hydrogen Bond Acceptor Group

Hydrogen bond acceptor groups present at the C-4 position of the phenyl ring of an alkoxy bridge improve the potency of a compound [131].[49]

131

Piperzinyl Group

Replacement of a triazole group with piperazine, as in compound [132], causes a complete loss of activity.[49]

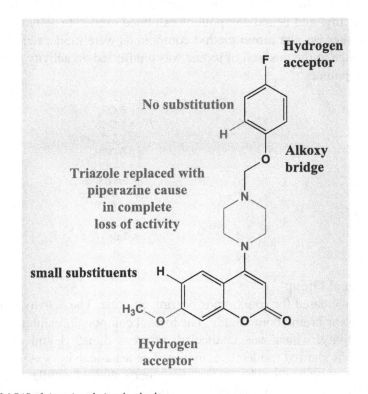

132

3.3.5.5 Iodinated Aryloxymethyl Coumarin

Various 4-iodophenol, 3-iodophenol, and 2-iodophenol coumarins were screened for two cancer cell lines (A-549 human lung carcinoma and MDAMB human adenocarcinoma mammary gland).[22]

Figure 3.6 SAR of piperazine substituted molecule.

2-idophenol Group

Mono-methyl and halogens groups on coumarins at 6- and 7-positions displayed effective activity, whereas the monomethyl compounds showed least activity.[22]

a = 6-CH$_3$
b = 7-CH3
c = 7,8-Dimethyl
d = 6-OCH$_3$
e = 7-OCH$_3$
f = 5,6-Benzo
g = 7,8-Benzo
h = 6-Cl
i = 7-Cl
j = 6-Br

22

3-idophenol Group

Bromo and chloro compounds exhibited potent increases in activity. The dimethyl and benzo compounds exhibited moderate activity in this series. Likewise, the mono methyl compounds were moderately active. The change in the position of iodine was unaffected on activity of methoxy compounds.[22]

a = 6-CH$_3$
b = 7-CH3
c = 7,8-Dimethyl
d = 6-OCH$_3$
e = 7-OCH$_3$
f = 5,6-Benzo
g = 7,8-Benzo
h = 6-Cl
i = 7-Cl
j = 6-Br

23

4-idophenol Group

Activity increased for compounds bearing chlorine. The activity remains the same for bromo compounds. The methyl compounds exhibited moderate activity, which was unaffected, whereas dimethyl and methoxy compounds showed moderate activities. The lease activity was exhibited by benzo compounds. The chloro compounds showed excellent activity with MIC 3.125 mg/mL and were equally potent to pyrizanamide.[22]

a = 6-CH$_3$
b = 7-CH3
c = 7,8-Dimethyl
d = 6-OCH$_3$
e = 7-OCH$_3$
f = 5,6-Benzo
g = 7,8-Benzo
h = 6-Cl
i = 7-Cl
j = 6-Br

24

3.3.6 Fused Ring at C$_3$-C$_4$ Position
3.3.6.1 Benzosubernone Ring

Benzosuberone-bearing coumarin moieties were evaluated for their cytotoxicity against four human cancer cell lines: HeLa (human cervical cancer cell line), A549 (human alveolar adenocarcinoma cell line), MCF-7 (human breast adenocarcinoma cell line) normal cell line HEK293 (human embryonic kidney cell line), and MDA-MB-231 (human breast adenocarcinoma cell line). Compound [133] with a benzene ring without any substitution displayed favorable antitumor effects, in contradiction to MCF-7 and A549 cell lines with IC50 values of 9.63 and 3.35 mM, respectively. Introduction of electron donating groups (-CH$_3$, -OCH$_3$) on benzene ring A was associated with a noticeable increase in the cytotoxic effect against all the four tested cell lines. Substitutions on ring A played an important role in exhibiting cytotoxicity due a conjugation effect on the coumarin moiety and were more favorable than having substitutions on ring D.[50]

133	R	R$_1$
a.	H	H
b.	CH$_3$	Cl
c.	OCH$_3$	H

133

3.3.6.2 Furan Ring

The furan ring is significant for cytotoxic activity. It was also revealed that a methyl furan ring resulted in better activity than an unsubstituted furan ring [134].[51]

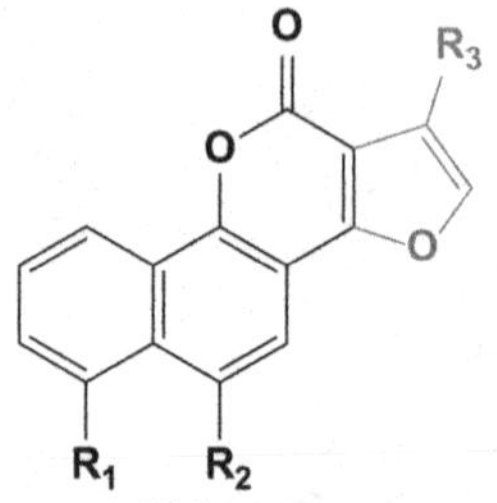

134

3.3.7 Functional Group at C_5 Position

3.3.7.1 Hydroxyl Group

5-hydroxy-6,7-methylenedioxycoumarin [135] in the tested concentration range was able to inhibit cell growth.[52]

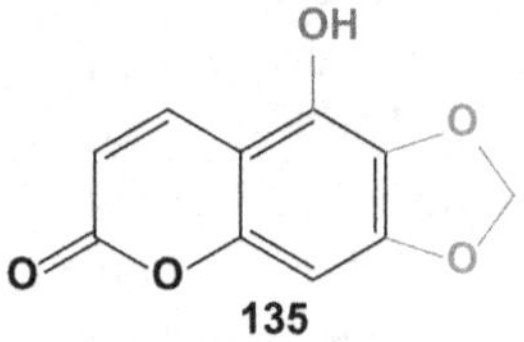

135

3.3.7.2 Alkoxy Group

The existence of an alkoxy group at the C-5 position of the coumarin ring enhanced proliferative activity; thus 5-oxygenated-6,7-methylene-dioxy-coumarins [136] and [137] showed differentiation and antiproliferative activity in U-937 cells.[52]

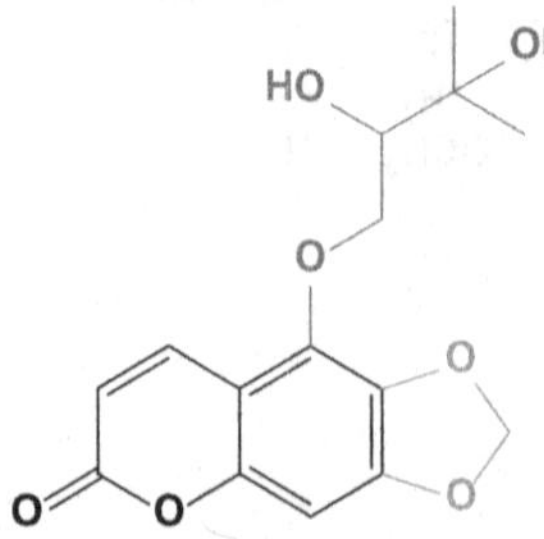

136

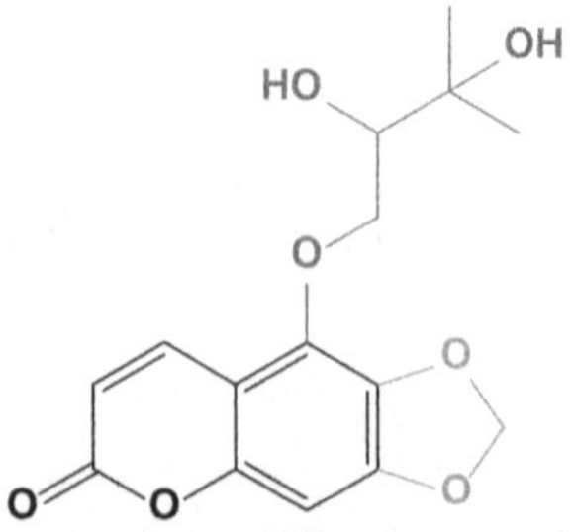

137

6,7-methylenedioxycoumarins without the alkoxy substituent at C-5 do not inhibit U-937 proliferation. Thus, compounds [138-141] does not exhibit any potency.[52]

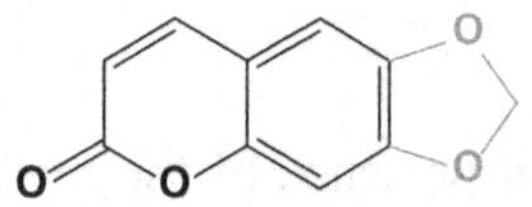

138

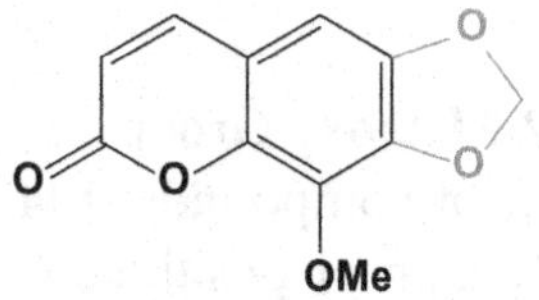

139

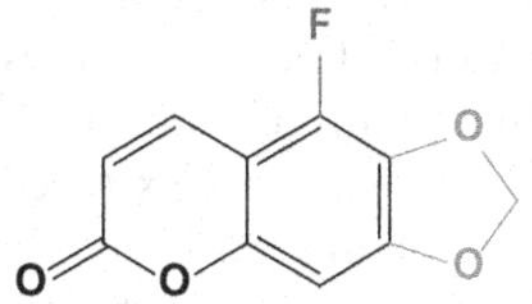

140

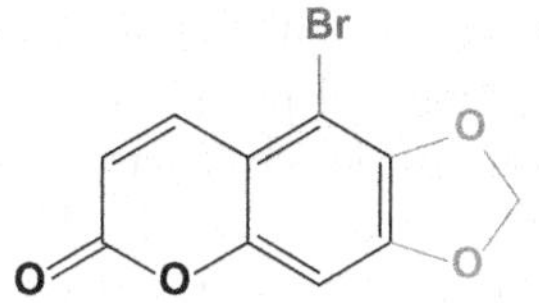

141

The compound having methoxy groups at carbon 5 and C-8 does not demonstrate an antiproliferative potency on leukemic cells, such as in compound [142].[52]

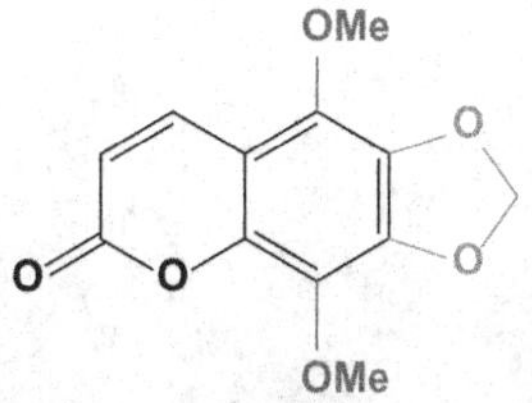

142

3.3.8 Functional Group at C$_6$ Position
3.3.8.1 Hydroxyl Group

Compound 6-hydroxycoumarin [143] was most fascinating; however, it is not an hCA I and hCA II inhibitor, yet it showed submicromolar inhibitory effects for isoforms hCA IX and hCA XII, with KIs of 0.198–0.683 lM.[41]

143

3.3.8.2 Methoxy Group

The set of compounds [144-146] that have a presence of the methylendioxy group at positions 6 and 7 of the nucleus supports the requirement for the differentiation activity.[52]

Mitogen-activated protein kinase (MAPK) are the main signaling cascades to transmit signals from their receptors to regulate gene expression of normal cell proliferation, survival, and differentiation. Abnormal regulation of MAPK cascades subsidize mainly to cancer. Extracellular signal-regulated kinase (ERK) is a component of this pathway that is stimulated by the rapidly accelerated fibrosarcoma (RAF) serine/threonine kinases. Mutational stimulation of RAF is responsible for the activation of oncogenesis and a key effector of the growth factor receptor (GFR). Thus, blocking the activity of the GFR-Ras-RAF-MEK-ERK signaling network is an effective way to attack tumors.

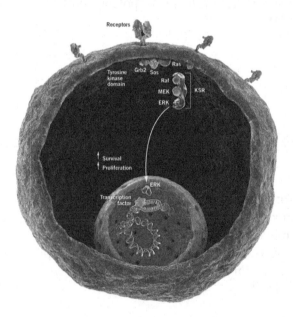

Figure 3.7 MEK pathway. www.biooncology.com/therapeutic-targets/mek.

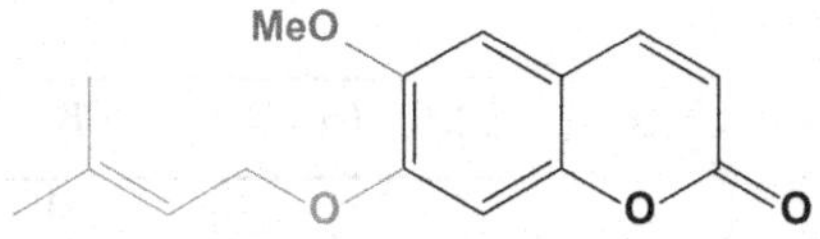

144

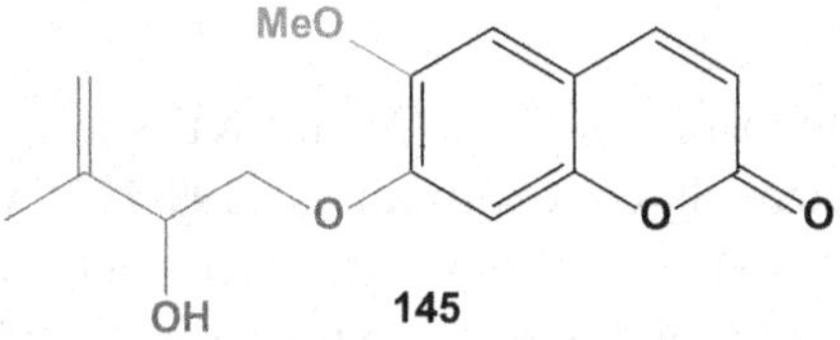

145

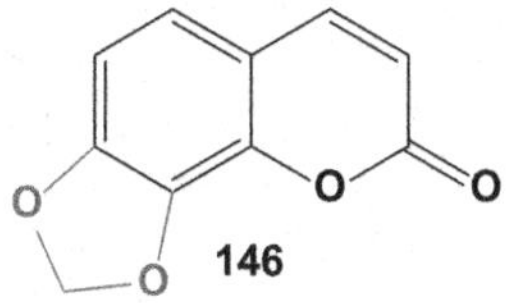

146

3.3.8.3 Chlorine Group

Widespread SAR studies were achieved for MEK1 inhibitory couma-rin derivatives (anticancerous and anti-inflammatory). Chlorine at position 6 of coumarin is one of the effective derivatives [147].[53]

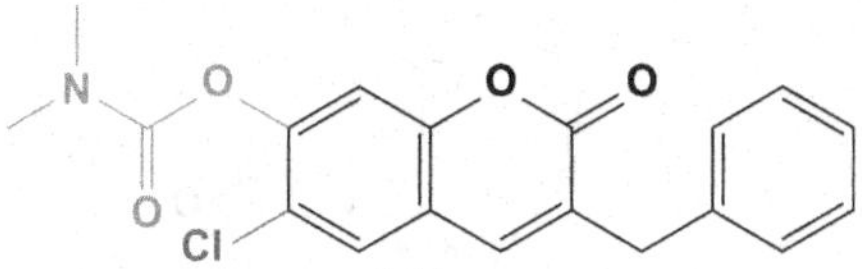

147

3.3.9 Functional Group at C_7 Position

3.3.9.1 Hydroxyl Group

In the present studies, in general the hydroxyl substitutions in the cou-marin molecule at C-7 [148a-e] demonstrated considerable cytotoxicity against the selected cancer cells.[40,41]

148	R$_1$	R$_2$	R$_3$	R$_4$	R$_5$	R$_6$
a.	H	H	H	H	OH	H
b.	H	CH$_3$	H	H	OH	OH
c.	H	H	CH$_3$	H	OH	H
d.	C$_6$H$_5$	CH$_3$	H	H	OH	H
e.	C$_6$H$_5$	OH	OH	H	OH	H

148

The tumor-related isoform hCA IX and XII was moderately inhibited by the simple hydroxylated compounds [149-150] with KI values of 478 − 560 nM and 754 − 8100 nM range, respectively. The 7-hydroxy coumarin [150] weakly inhibited the cytosolic isoform hCA I (h, human enzyme).[44]

149

150

3.3.9.2 Chlorine Group

The most effective hCA IX inhibitor was 7-chloro-4-chloromethyl-coumarin, with KIs of 0.198−0.359 μM.[41]

151

3.3.9.3 Alkoxy Group

Methoxy Group

The set of compounds [152-154] having presence of the methylendioxy group at positions 6 and 7 of the nucleus supports the requirement for the differentiation activity.[52]

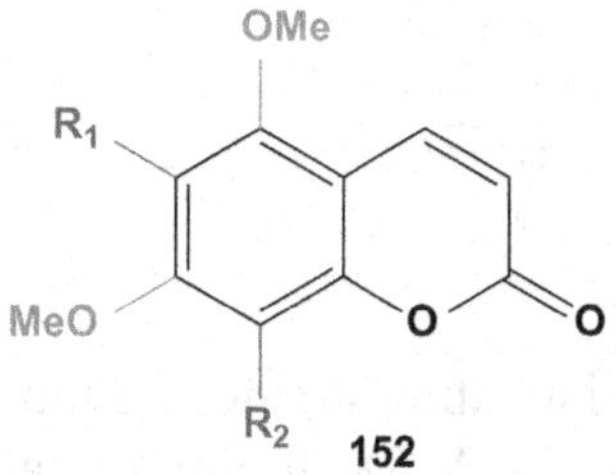

152	R$_1$	R$_2$
a.	H	OH(leptodactylone)
b.	H	OMe
c.	OH	H (fraxinol)

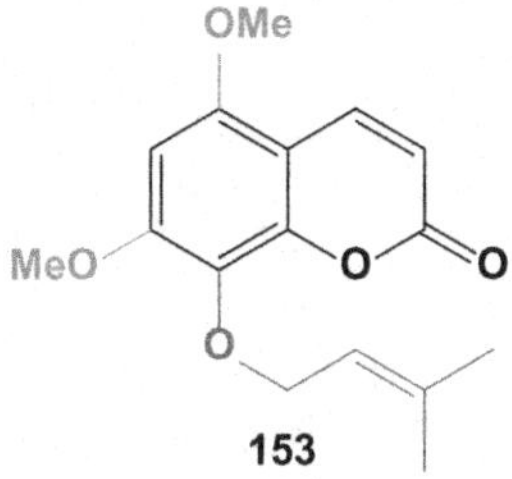

153
artanin

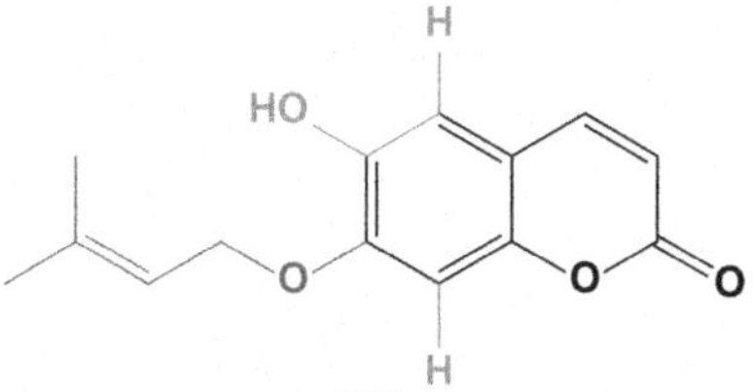

154

7-substituted derivatives were found to be improved inhibitors compared to the corresponding 6- or 4-substituted derivatives. Compounds [155-157] displayed significant submicromolar hCA IX inhibition. Among them, tosylate [156] was the greatest inhibitor (KI of 0.53 lM), followed by fluoro derivative [157] (KI of 0.68 lM), whereas slightly weaker inhibition activity was shown by hydroxyethyl derivative [155] (KI of 0.92 lM).[43]

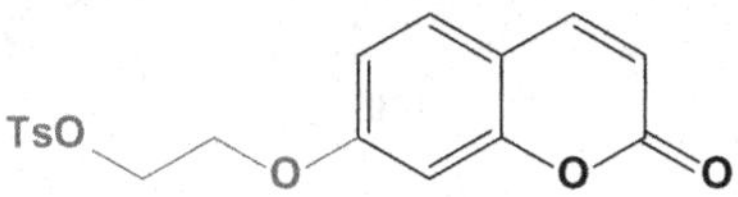

156

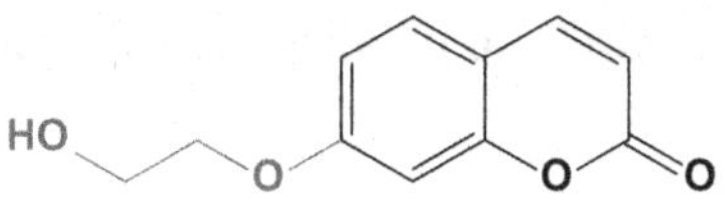

155

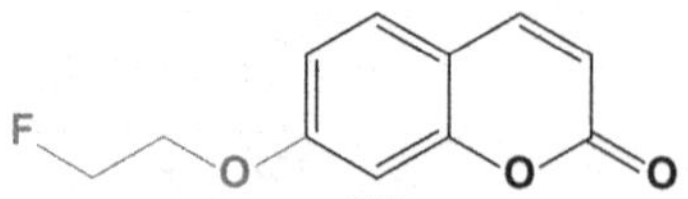

157

3.3.9.4 Amide Group

Amino-coumarin derivatives [158-160], including ureido, acetamido or carbamate moiety, displayed enhanced hCA IX inhibition activity, compared to unsubstituted aminocoumarin.[43]

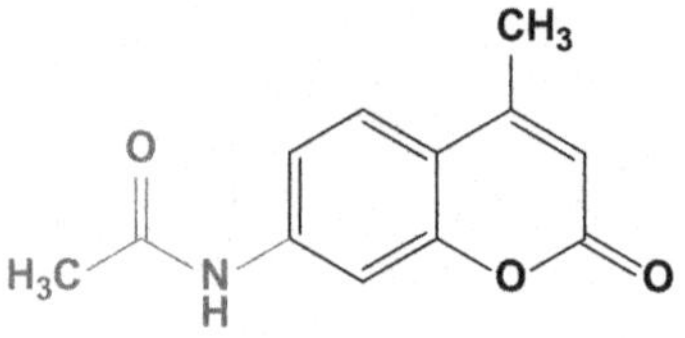

158

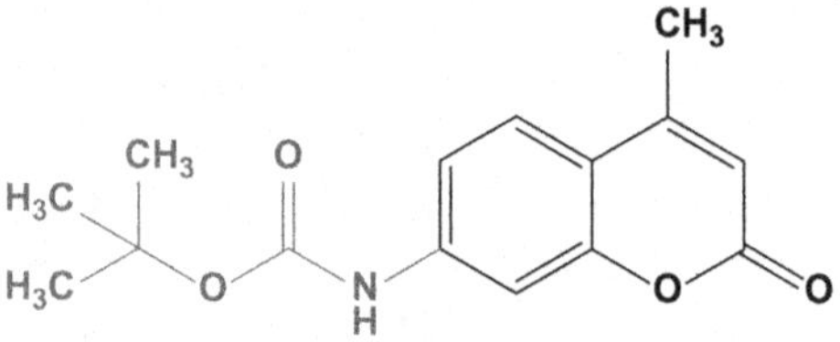

159

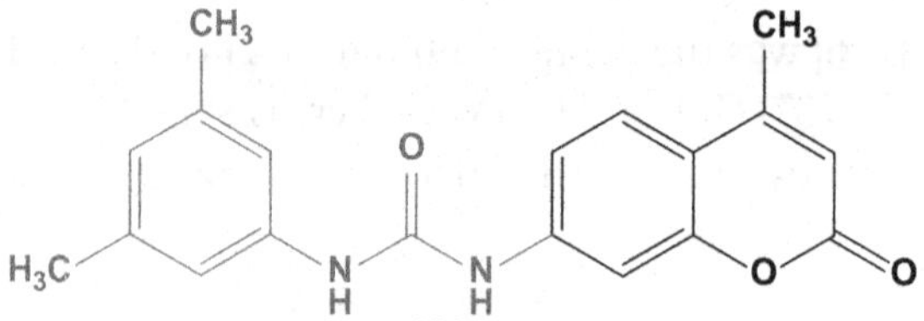

160

3.3.9.5 Sulfonate Group

The aromatic sulfonates [161-162] are more favorable for antiproliferative activities than are aliphatic or alicyclic sulfonates.[54]

161	R
a.	Me
b.	Et
c.	n-Pr
d.	Cyclo-Pr
e.	Ph
f.	p-Tolyl
g.	p-(CF$_3$)C$_6$H$_4$
h.	p-(tert-butyl)C$_6$H$_4$
i.	p-(F)C$_6$H$_4$

162	R
a.	Me
b.	Et
c.	n-Pr
d.	Cyclo-Pr
e.	Ph
f.	p-Tolyl
g.	p-(CF$_3$)C$_6$H$_4$
h.	p-(tert-butyl)C$_6$H$_4$
i.	p-(F)C$_6$H$_4$

3.3.9.6 Sugar Derivatives

7-substituted coumarins were examined as inhibitors of zinc-containing carbonic anhydrase enzymes. These sugar-substituted coumarins were weak inhibitors mainly for isoforms CA I and II; however, a few of them are effective in inhibiting the tumor-associated CA IX and XII, even in the low nano-molar range. A weak inhibition of isoform hCA II was obtained by all sugar-substituted coumarins with KI = 77 − 100 μM. Coumarin derivatives substituted with either mannose, rhamnose, or ribose [163-165] showed weak inhibition of cytosolic isoform hCA I within the range of 58.4 − 100 μM.[44]

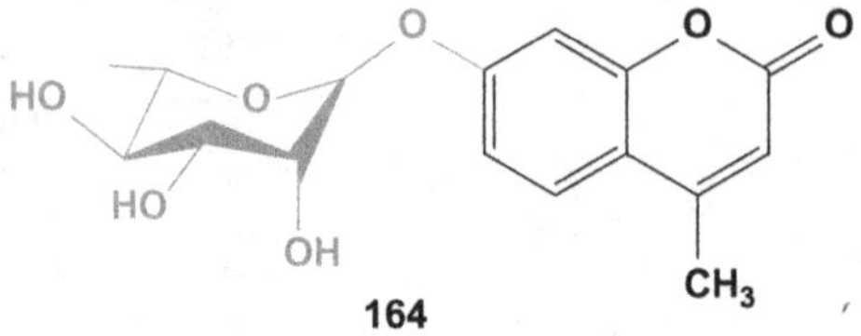

164

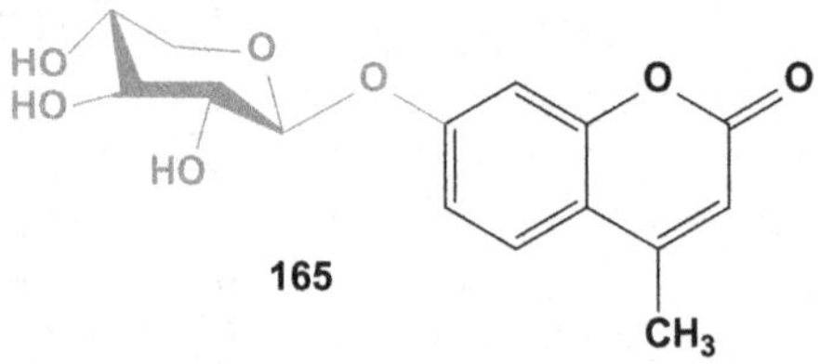

165

Small dissimilarities in the coumarin structure intensely effected the inhibitory activity of the compound because the hydroxyl-cinnamic acids formed after hydrolysis can achieve cis or trans conformations and relate with numerous amino acid residues at the entrance of the enzyme active site. The glucose, xylose, galactose, and melobiose coumarin derivatives [166-168] were shown medium potency as hCA I inhibitors (with KI values in the range of 0.59 − 3.4 μM).[44]

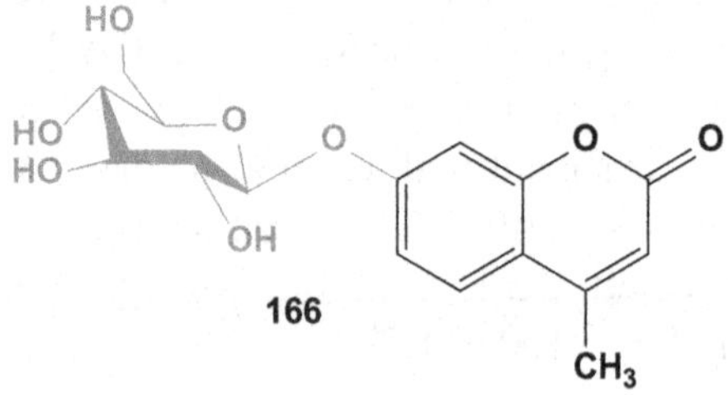

166

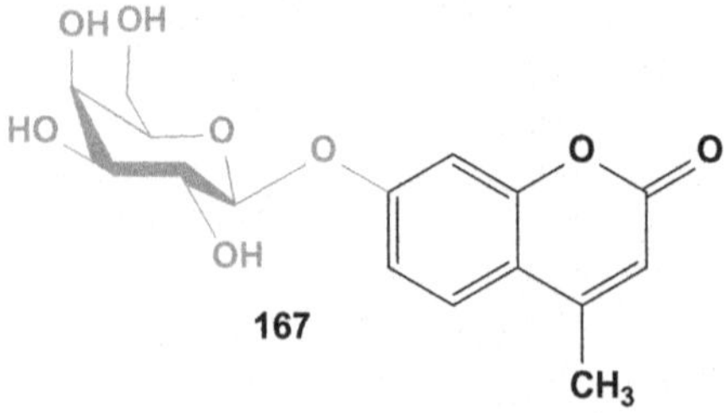

167

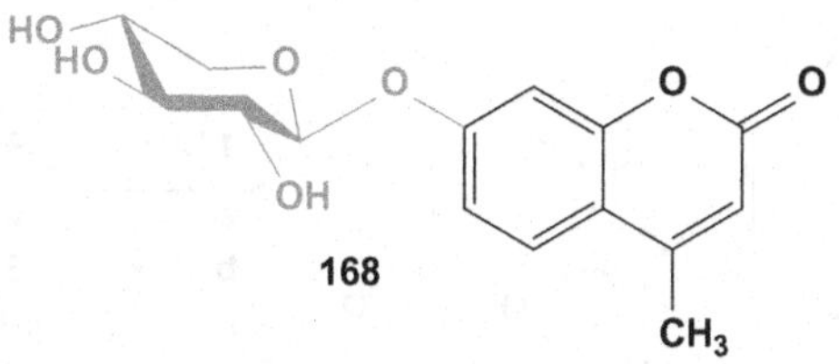

168

The hCA II inhibition was displayed by disaccharide derivatives of glycosyl coumarins [169] with KI = 0.59 μM.[44]

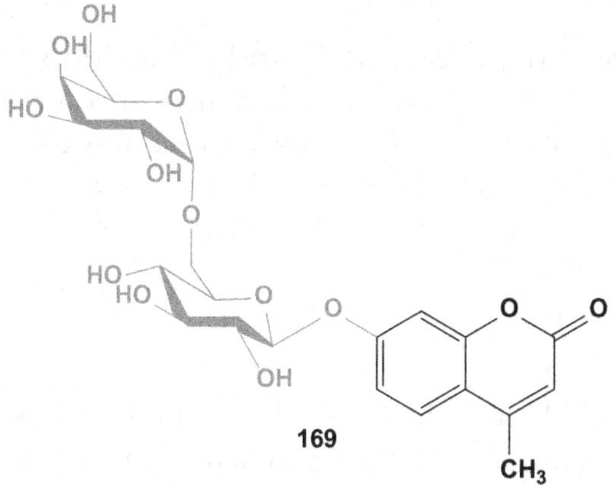

169

3.3.10 Disubstituted Group at C_6 and C_7 position

6, 7-dihydroxycoumarin derivatives were well known as a lead molecule due to having tumor cell-specific cytotoxicity.[41] The 6, 7-disubstituted coumarin moieties [170, 171] and [172a-f] exhibited weak or complete absence of CA inhibition effects compared to all examined isoforms.[52]

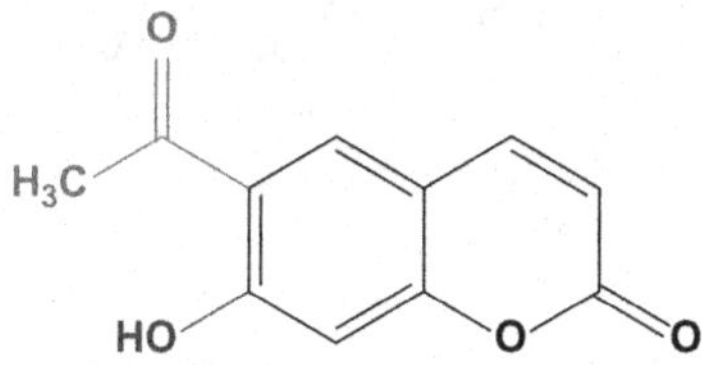

170

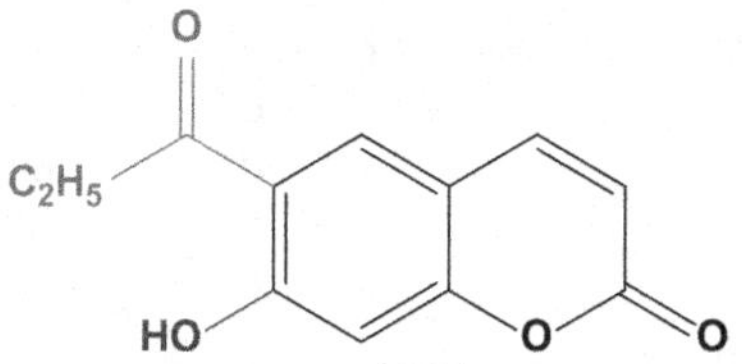

171

172	R
a.	Me
b.	Et
c.	n-Pr
d.	n-Bu
e.	Bn
f.	

172

3.3.11 Disubstituted Group at C_7 and C_8 position

Coumarin rings with 7, 8-substituented derivatives were exhibited a totally diverse inhibition profile. Umbelliferone was effective in inhibiting the tumor-related hCA IX, but its derivatives were far better inhibitors. It was found that the activity enhanced from the acetyl to the propionyl derivatives.[55]

3.3.11.1 N-Propyl Group

In the ethers [173,174a-f] the n-propyl derivative was the best hCA IX inhibitor, then decreased for the benzyl and adamantly ethyl derivatives, proving that SAR is less sharp for these two derivatives conflicted by the presence of such a bulky group. It was also found that the propionyl derivatives [175,176a-c] are more effective related to the analogous acetyl derivatives.[55]

173

174

174	R
a.	Me
b.	Et
c.	n-Pr
d.	n-Bu
e.	Bn
f.	

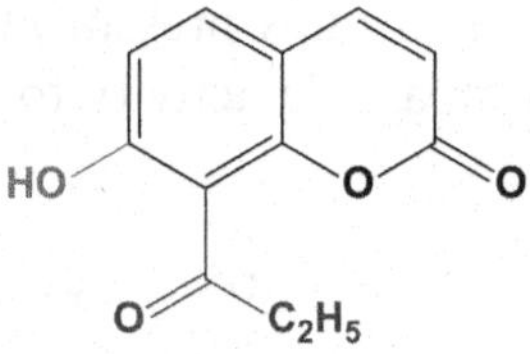

175

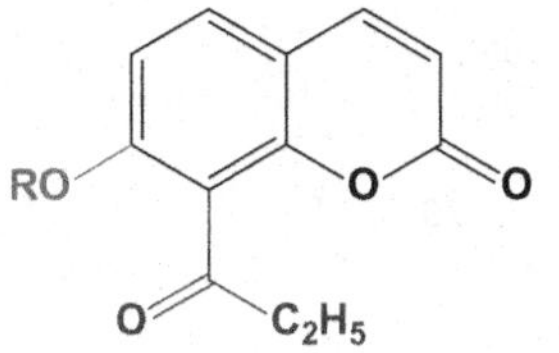

176

176	R
a.	Me
b.	Et
c.	n-Pr

3.3.12 Fused Ring at C_7-C_8 Position

3.3.12.1 Benzene Group

Methyl and ethyl moiety at the C-4 position of the benzene ring [177-178] may enhance the cytotoxic effect intensely.[51]

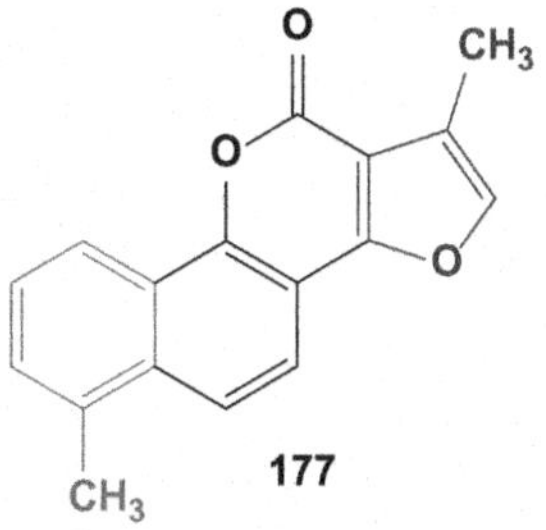

177

Neo-tanshinlactone

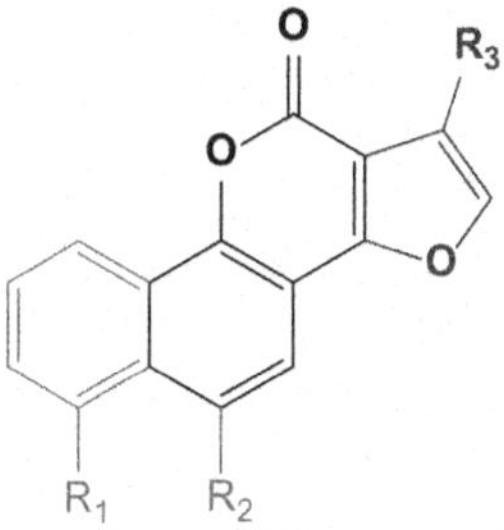

178

R_1 = CH_3, C_2H_5 or H
R_2 = H, Cl or OCH_3
R_3 = CH_3 or OH

Compounds with bulkier groups present at the C-4 position of the benzene ring [179] had comparable activity to the lead compound neo-tanshinlactone [177].[51]

179

CHAPTER 4

Neuroprotective Agents

Archi Sharma

Department of Chemistry, National Institute of Technology, Raipur, Chhattisgarh, India

4.1 INTRODUCTION

Neuroprotective agents are used to save ischemic neurons in the brain from irreversible injury or offer protection against cell degeneration to the neuronal cells. An ischemic neuron occurs when a cerebral vessel occludes, obstructing blood flow to a portion of the brain. Neuroprotective agents are thought to stop nerve cells dying. The term *neuroprotective* refers to the relative preservation of neuronal structure and/or function. Several neuroprotective agents are used to treat other conditions as well. Drugs such as anti-Alzheimer's, anti-Parkinson's, and anti-ischemics[56] are neuro-protective agents.

The *N-methyl-D-aspartate receptor (NMDAR)*, a glutamate receptor, is the main molecular part for controlling memory function and synaptic plasticity. An NMDA receptor is characterized by its voltage-dependent activation, which is due to the outcome of an ion channel block by extra-cellular Mg^{2+} & Zn^{2+} ions. This permits the movement of Na^+ and a small quantity of Ca^{2+} ions inside the cell and K^+ ion outside the cell. An antibody encephalitis acts as an anti-NMDA receptor, which is effectively harmful; however, it has great possibility for recovery. It is caused by an autoimmune response, primarily against the NR1 subunit of the NMDA receptor.

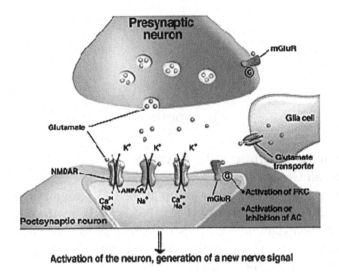

Figure 4.1 NMDAR pathway. http://pubs.niaaa.nih.gov/publications/arh314/310-339.htm.

4.1.1 Functional Group at C_4 Position

4.1.1.1 Methyl Group

4-methylsubstituted derivative UBP714 [**180**] showed higher effective activity at recombinant NMDARs having GluN2A or GluN2B and smaller activity for the GluN2D subunits with receptors. At the C-4 position of the coumarin ring, require a hydrophobic group for probable effect at the binding site.[57]

UBP714
180

4.1.2 Fused Ring at C_5-C_6 Position

A fused phenyl ring at C-5 and C-6 positions of the coumarin ring provides UBP659 [181]. This compound showed the weaker NMDAR antagonist activity, signifying that the substituted group at the C-5 position is not applicable for binding the NMDAR at its active sites.[57]

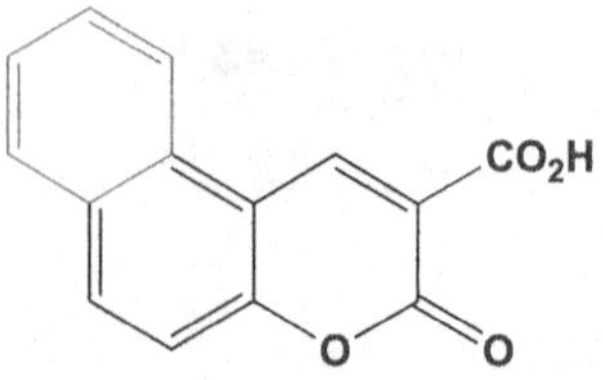

UBP659
181

4.1.3 Functional Group at C$_7$ Position
4.1.3.1 Small Group
The presence of an extra polar group at the C-7 position of the coumarin ring—for example, hydroxyl (UBP652) [183], methoxy (UBP653)
[184], or diethylamino (UBP654) [185] groups—did not improve
N-methyl-D-aspartate receptor antagonist activity, compared to the
parent compound UBP649 [182].[57]

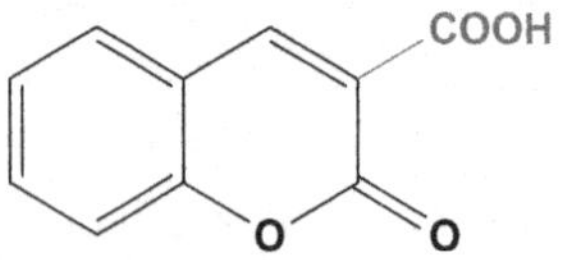

UB9649
182

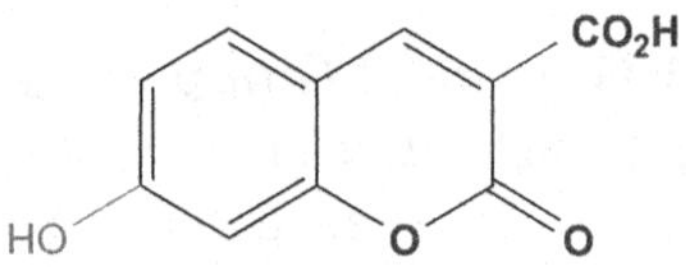

UBP652
183

4.1.3.2 Methoxy Group [184][57]

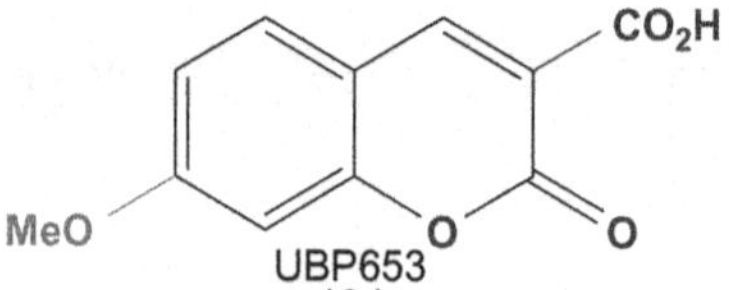

UBP653
184

4.1.3.3 Diethylamino Group [185][57]

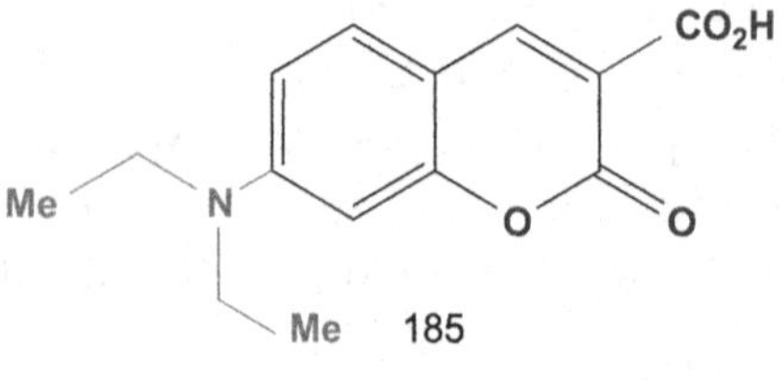

185

4.1.4 Fused Ring at C$_6$-C$_7$ Position

4.1.4.1 Furanocoumarins

Dihydro-furanocoumarins [186–190], furanocoumarin and isoimperator-in [191] also expressively diminished the glutamate-induced toxicity, exhibiting cell feasibilities. Cyclization at the C-6 position of coumarin of isoprenyl components, such as dihydrofuran or dihydropyran, or the furan ring is imperative in the neuroprotective action of coumarins.[58]

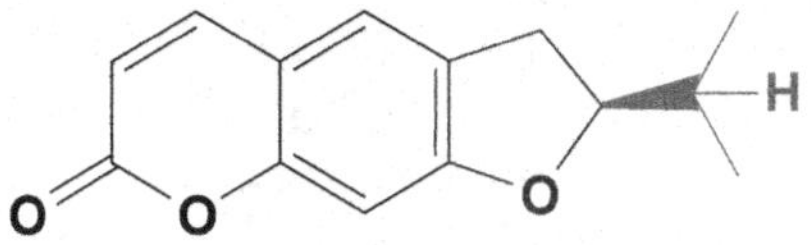

186

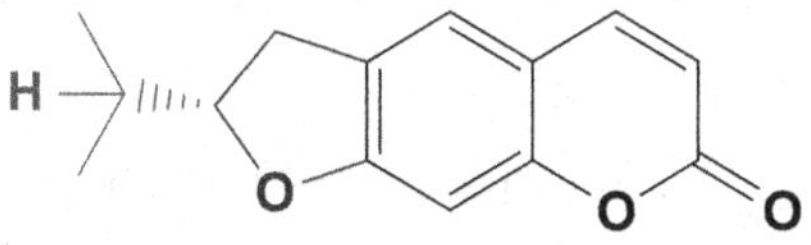

187

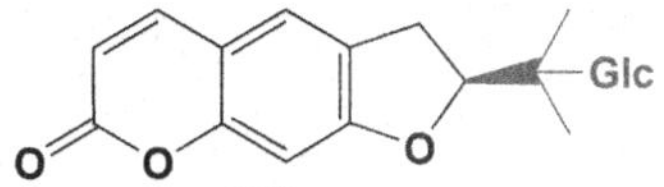

188

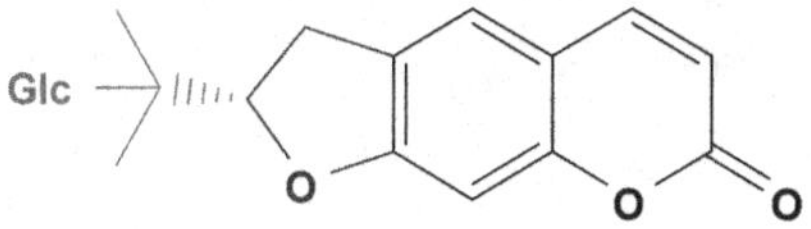

189

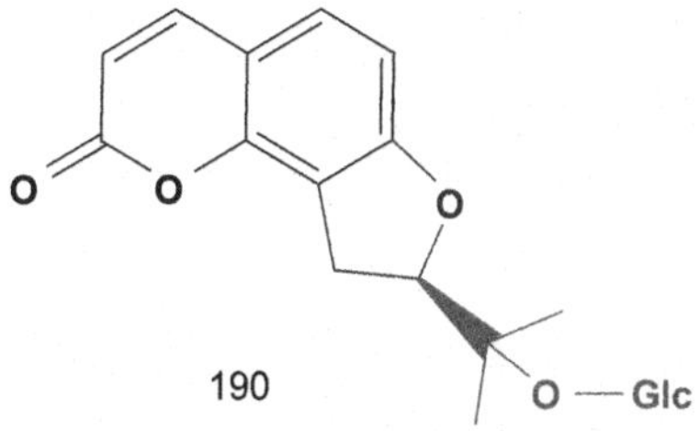

190

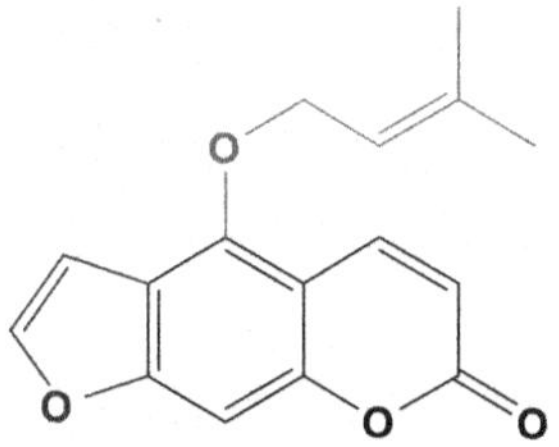

191

Umbelliferone [149] displayed important neuroprotective activity, on the other hand; its glycosides [192], apiosyl skimmin and skimming [193], do not exhibit any significant activities. These outcomes recommended that the improved hydrophilicity owing to addition of a sugar moiety can decrease neuroprotective activity.[58]

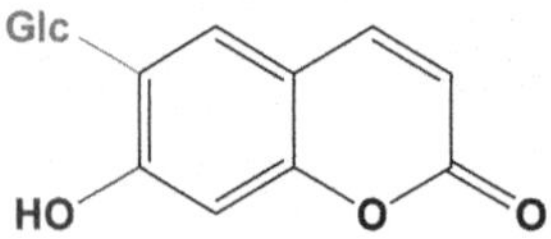

192

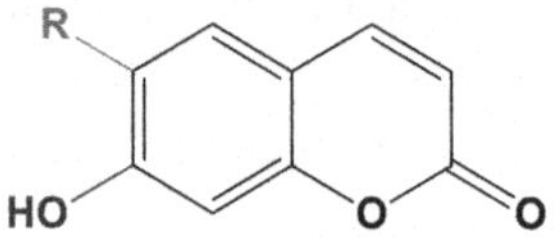

R = Api(1"-6")Glc

193

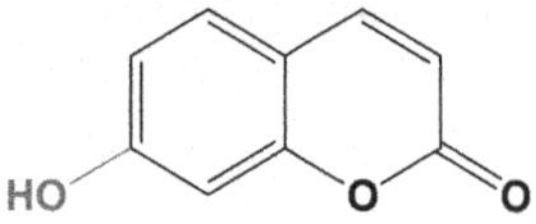

149

4.1.4.2 Pyranocoumarins

Pyranocoumarins exhibited important neuroprotective actions.[58]

Dihydropyran group
Dihydropyran moiety of decursinol [194–198] can play an essential role in neuroprotective activity, except xanthyletin [194].[58]

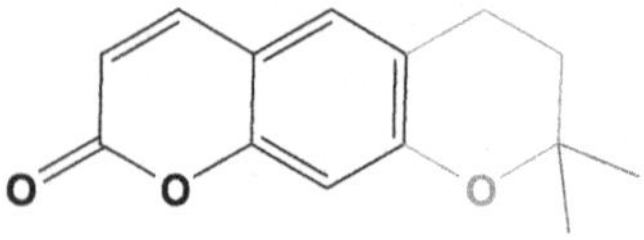

xanthyletin
194

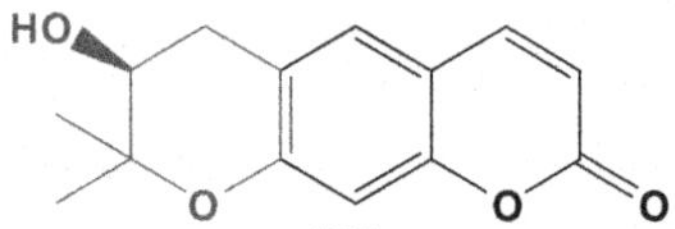

195

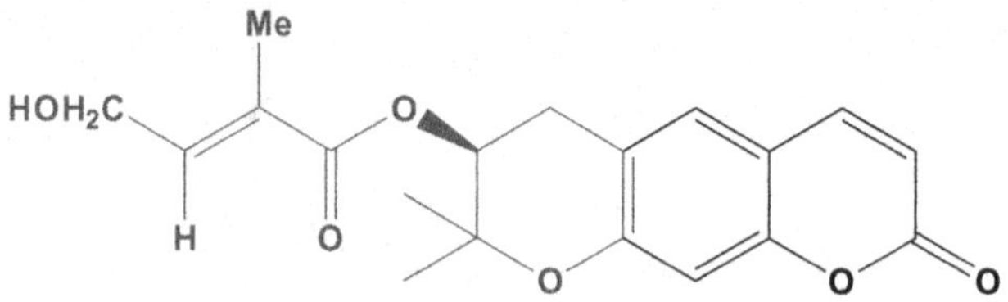

196

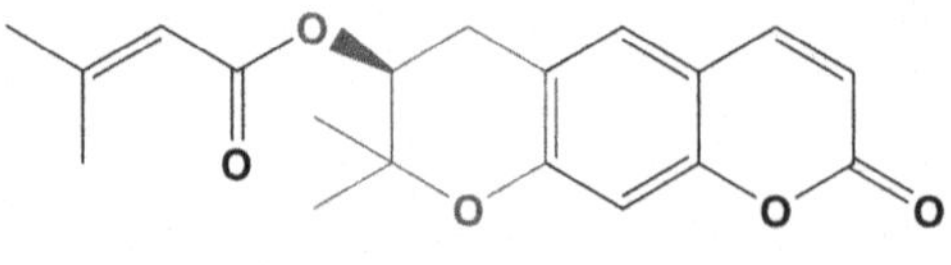

197

198

Epoxy Group

Epoxy angeloyldecursinols and decursinol have effective neuroprotective activities equivalent to that of MK-801. An epoxy-substituted coumarin [199] performed more effectively than those without epoxide functionality, signifying that the epoxidated hydroxyl isoprenyl group may improve the neuroprotective action of coumarins.[58]

199

a.(2S", 3S")

b.(2R", 3R")

$$R = $$

4.1.5 Functional Group at C_8 Position

4.1.5.1 Bromine Group

A bromine atom at the C-8 position of UBP608 enhances the NMDAR antagonist activity, and it decreases selectivity for GluN2A. 6,8-dibromo-(UBP651) substituted coumarins [201] showed better NMDAR antagonistic activity compared to the mono-substituted bromine at the C-6 position (UBP608) [200].[57]

200

201

4.1.5.2 Iodine Group

Substitution of the large group, such as iodo at the C-8 position on the coumarin ring, enhances N-methyl-D-aspartate receptors (NMDAR) antagonist activity. Compounds (UBP658) substituted with di-iodine at C-6 and C-8 positions of the coumarin [203] have better NMDAR antagonistic activity than the mono-substituted 6-iodo (UBP657) derivatives [202].[57]

I—[structure] CO_2H
202

I—[structure] CO_2H
203
I

4.2 ANTI-ALZHEIMER AGENTS

4.2.1 Introduction

Alzheimer's disease (AD) is the most common type of dementia. There is no effective remedy for the disease, which degrades the neural function during its advancement and ultimately leads to death (Figure 4.2). Most frequently, AD is diagnosed in people over 65 years of age, whereas the early-onset variety of Alzheimer's can occur much earlier.[59]

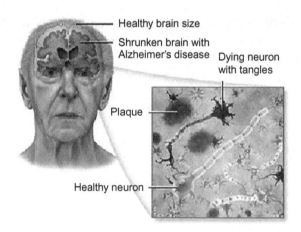

Figure 4.2 Alzheimer's disease. www.buzzoop.com/health/5-awesome-health-benefits-eating-apple/.

Acetylcholine is a neurotransmitter (a brain chemical) that helps in think-ing and memorizing. In Alzheimer's disease, acetylcholine becomes degraded. Thus, Alzheimer's victims gradually make less acetylcholine. These two phenomena cause the continuing damage of memory and thinking ability. Medicines called *cholinesterase inhibitors* were meant to stop acetylcholine from degrading. They may help the brain cells work better. However, it is difficult to stop or reverse the damage to brain cells and degradation of acetylcholine that occur in Alzheimer's disease. Hence, these cholinesterase inhibitors cannot stop the illness from getting worse, though the process can be slowed. These treatments do not help in producing acetylcholine. Therefore, after a period the treatments may stop working.

4.2.2 Unsubstituted Coumarin

Coumarin without substitutions on the benzene ring displayed no inhibitory effect on BACE1.[60]

4.2.3 Carbon Chain Length

AChE inhiitory effect of the omega-[*N*-methyl-*N*-(3-alkylcarbamoylox-yphenyl)methyl]aminoalkoxyxanthen-9-ones were studied. It was found that the compound associated with three-carbon chain length showed the best inhibitory activity.[61]

4.2.4 Alkyl Group

The coumarin derivatives that contain methyl, *n*-butyl, or *n*-heptyl sub-stituents [204–206] are equally potent.[61]

Methyl group

204

n-Butyl group

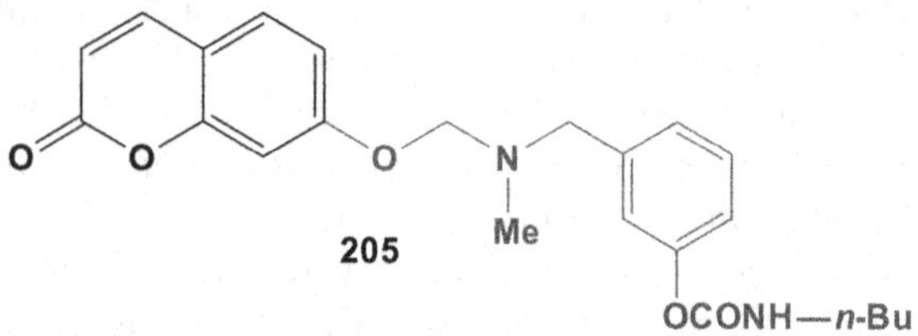

205

OCONH—*n*-Bu

n-Heptyl group

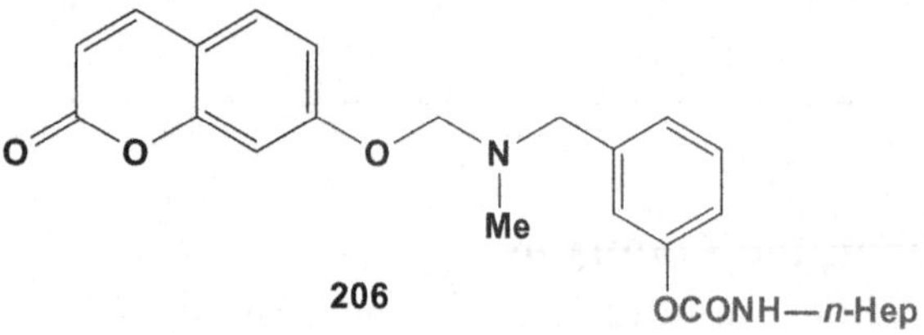

206

OCONH—*n*-Hep

4.2.5 Steric and Electronic Effect

A set of benzyl pyridinium substituted coumarin derivatives was evaluated for acetylcholinesterase (AChE) and butyrylcholinesterase (BuChE) inhibitors by using donepezil hydrochloride as standard drug. Anticholinesterase activity was largely influenced by the steric and electronic structures of the substituents. The movement of the fluorine group from position 2 [207] to position 4 [208] led to fourfold reduction in the activity.[62]

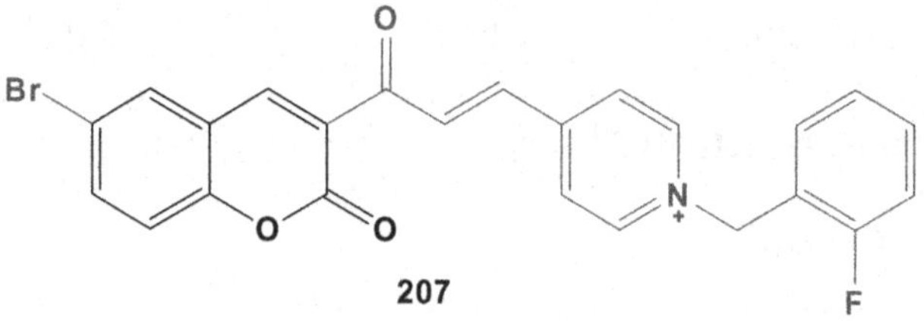

207

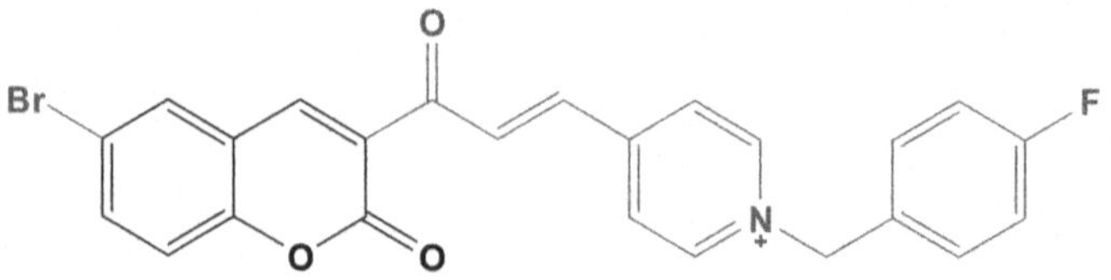

208

On the basis of docking studies, existence of fluoro at the ortho position, as in compound [207, 209], can interrupt the π-π interactions through rotation of the phenyl ring.[62]

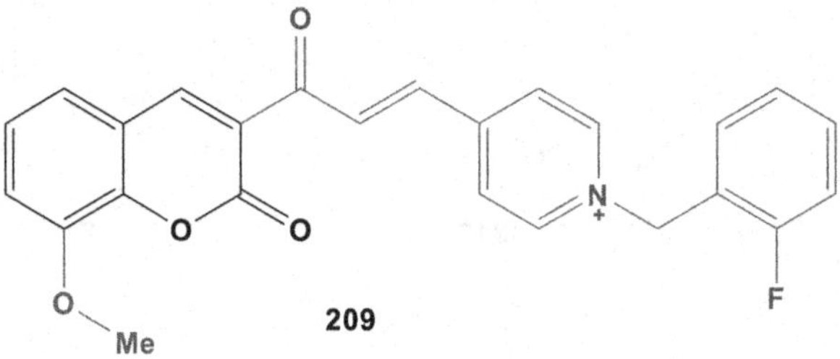

209

Fluorine at the meta position [210] consequently caused improved activity due to appropriate attachment of the phenyl ring. This phenomenon was expected due to the steric hindrance.[62]

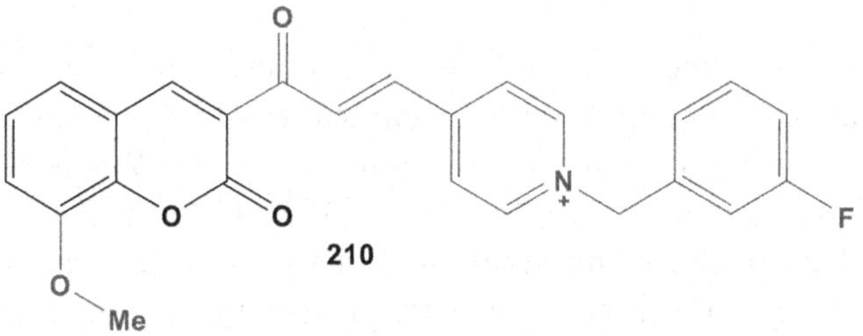

210

The size of para substituents intensely affects the activity of compounds. Therefore, bulky groups at the para position showed weaker activity, as compound [211–212].[62]

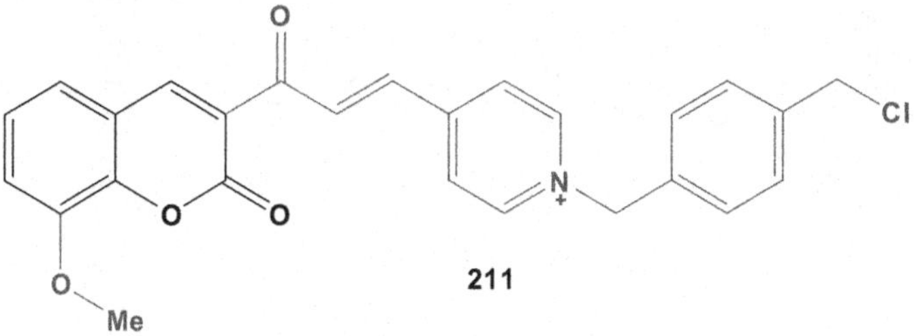

211

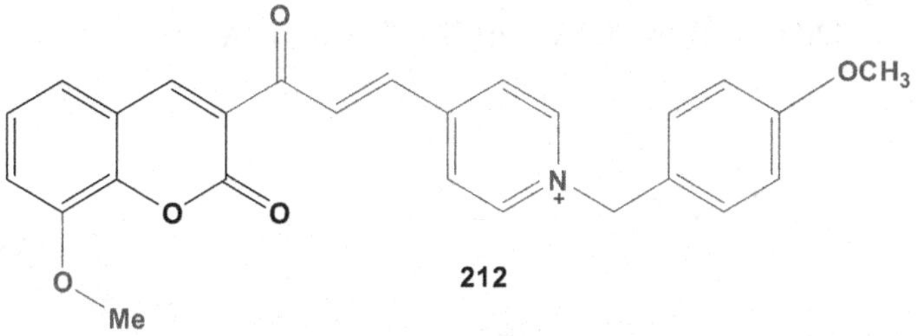

212

Elimination of the 6-bromine group from the coumarin ring [213–214] caused the large decrease in activity.[62]

213

214

At position 2 (meta position of the aromatic ring), the size of the substituent is important to show significant activity regardless of the electronic properties. Increasing the size of van der Waals radii led to increased anticholinesterase activity: $CH_3 > Cl > F$; 200, 180, and 135 pm, respectively. Compounds with a further electron-withdrawing group at the C-3 position of the benzyl nucleus demonstrated greater activity in the order of compounds [215] (F, 0.47 nm) > (CN, 76 nm) > (Cl, 86 nm) > (CH_3, 800 nm).[62]

215

4.2.6 Enzymatic Selectivity

Compound [204] is more active on BuChE than AChE.[61]

204

4.2.7 Functional Group at C$_3$ Position
4.2.7.1 Benzyl Alkyl Amino Group

The benzene ring with a *N*-benzyl-*N*-methyl group in its meta position, substituted with the coumarin ring, exhibited a lesser AChE inhibitory effect related to the lead compound [216]. A similar trend was revealed for mono methoxy derivatives, where the meta analogs had a greater IC50 than the para analogs.[63]

216

217

217	Chain	R$_1$	R$_2$
a.	meta	OCH$_3$	OCH$_3$
b.	para	OCH$_3$	H
c.	meta	OCH$_3$	H
d.	para	H	OCH$_3$
e.	meta	H	OCH3

4.2.7.2 Terminal Phenyl Group

A nitro group at the ortho, meta, and para positions, as well as -methoxy and -methyl derivatives [218a-i], were synthesized, showing a reduced anticholinesterase activity. Compounds substituted with *o*-methoxy groups showed an IC50 that is of the same command of scale as the lead compound, considering that the *o*-methoxy is able to balance the net-positive charge of the amino group through inductive and mesomeric effect.[63]

218

218	R
a.	*o*-NO$_2$
b.	*m*-NO$_2$
c.	*p*-NO$_2$
d.	*o*-CH$_3$
e.	*m*-CH$_3$
f.	*p*-CH$_3$
g.	*o*-OCH$_3$
h.	*m*-OCH$_3$
i.	*p*-OCH$_3$

4.2.7.3 Substituted Basic Nitrogen Group

The methyl ammonium analog [219] was synthesized to check the prominence of the positive charge, which displayed the lowest-activity IC50 of the series.[63]

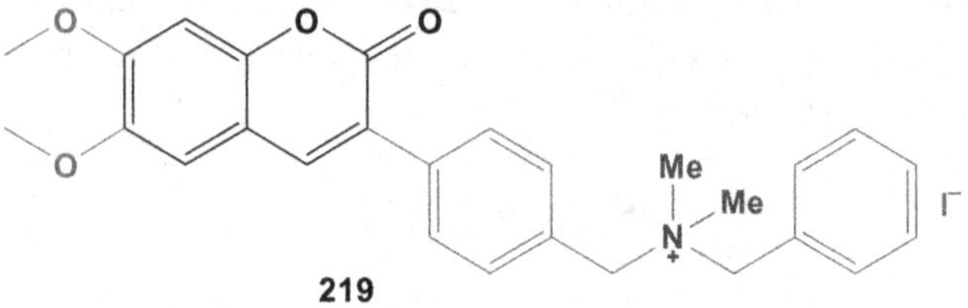

219

In replacing methyl with ethyl moiety of the substituted amine group [220], compared with the lead compound, it was found that an ethyl substituent exhibited a higher potency due to its better lipophilic effect of the compound. In the hydroxyl ethyl group [221] the increased lipophilic effect due to the ethyl group is compensated by the lower lipophilicity, developed by the presence of the OH group.[63]

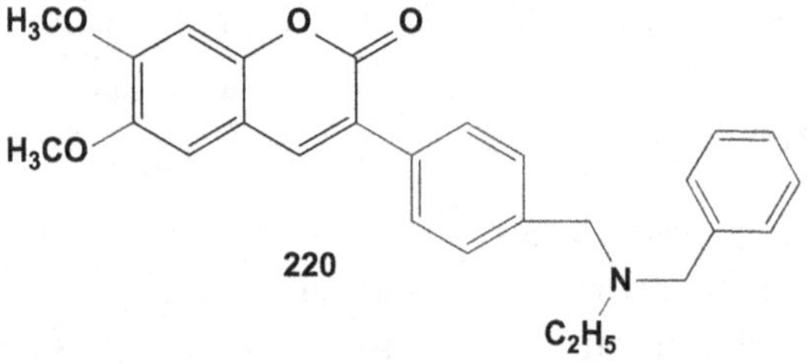

220

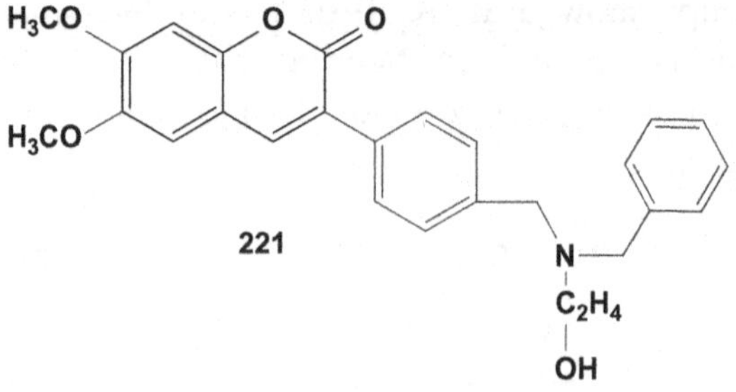

221

4.2.7.4 Heteroaryl Oxygenated Group

The inverted position of the phenyl ring and coumarin as shown in the compound [222] were found insoluble, thus the biological effect of this compound was difficult to measure.[63]

222

4.2.7.5 Biarylpiperazine Group

Biarylpiperazine containing coumarinyl moieties were evaluated as β-secretase inhibitors for their potential in Alzheimer's disease treatment. The compound was reported as one of the most potent, hence taken as a reference compound. Compounds [223], bearing various groups at the N_4-position as piperazinyl, were found almost equipotent, compared with the reference compound, whereas substituents such as Boc or oxygen at the N_4-position results in a reduction of the inhibitory activity.[64]

223

223	R_1	R_2	X
a.	H	Br	O
b.	H	Br	N-Boc
c.	H	Br	NH-TFA
d.	OMe	Br	O
e.	OMe	Br	N-Boc
f.	OMe	Br	N-benzyl
g.	OMe	Br	NH-TFA
h.	OMe	Br	NCH_2CH_2OH
i.	H	Ph	N-Boc
j.	H	Ph	NH-TFA

Substitution of hydroxyl ethyl [224] and benzyl [225] groups at the N_4-position increases the inhibitory activity.[64]

224

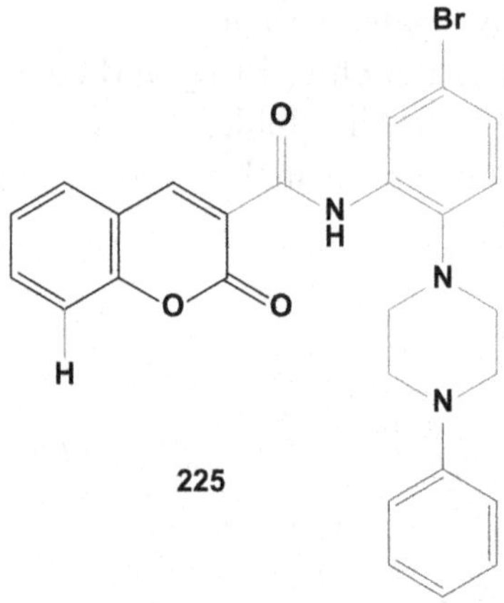

225

The presence of a biphenyl moiety [226] does not have a prominent effect on the inhibitory action of the subsequent analogs.[64]

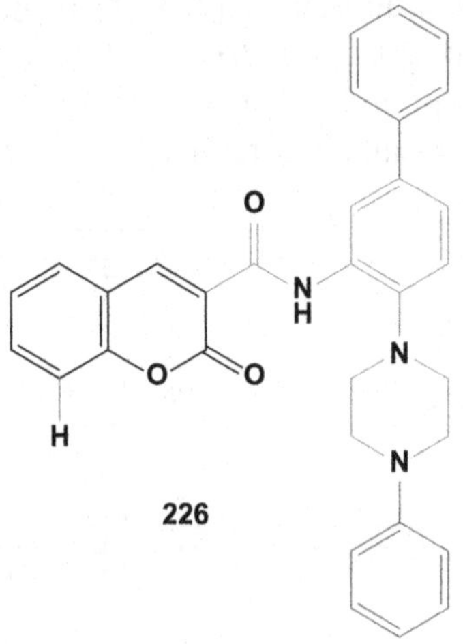

226

4.2.8 Functional Group at C_6 Position
4.2.8.1 Nitrogenous Group

With an electron-donating (NH_2) [227] and electron-withdrawing (NO_2) [228] substituent at the C-6 position of the coumarin ring with the *N*-benzyl-*N*-methyl substituents in the para position, both the derivatives showed significant anticholinesterase activity.[63]

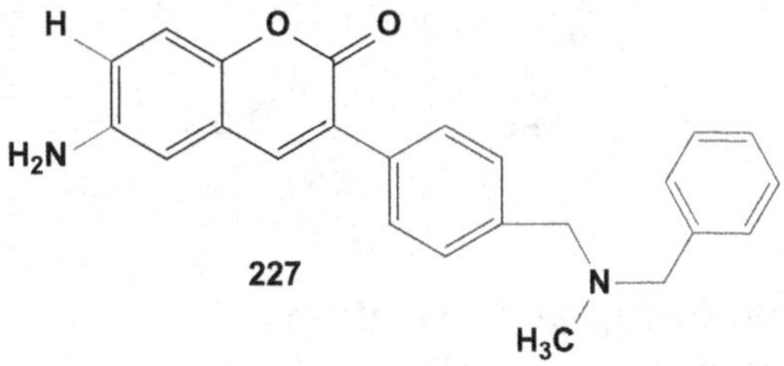

227

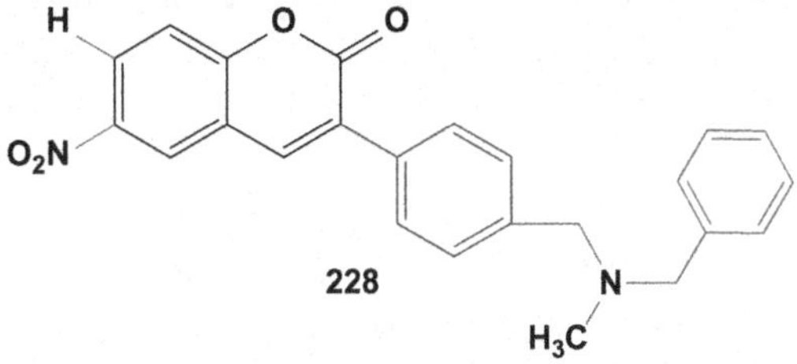

228

4.2.8.2 Isoprenyl Group

A noncyclized isoprenyl unit at C-6 of simple coumarins [umbelliferone, **229–230**] exhibits much-reduced inhibiting activity for AChE compared with the cyclized isoprenyl unit [**231–232**].[65]

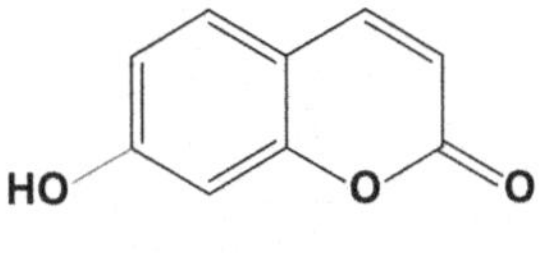

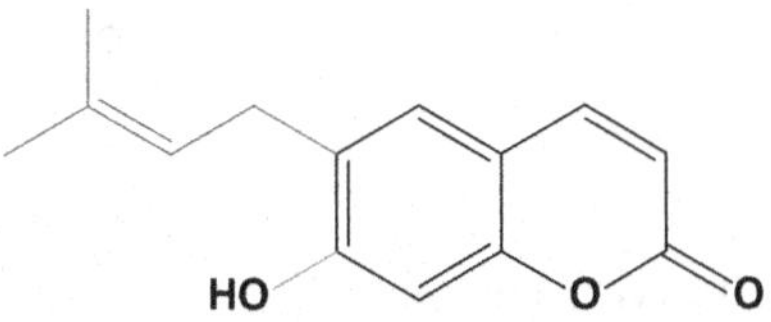

229

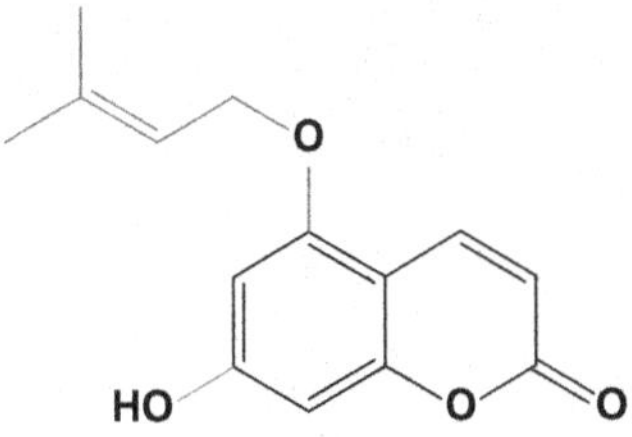

230

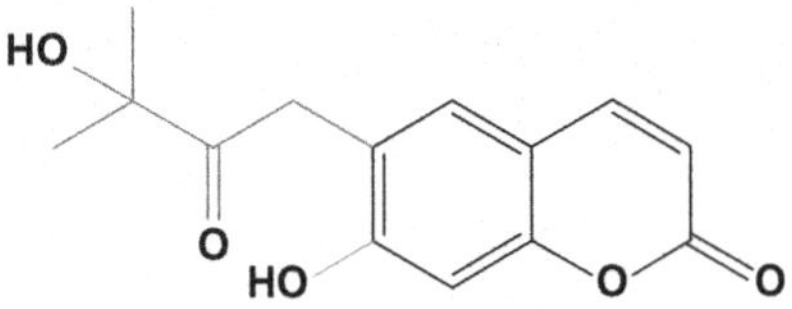

231

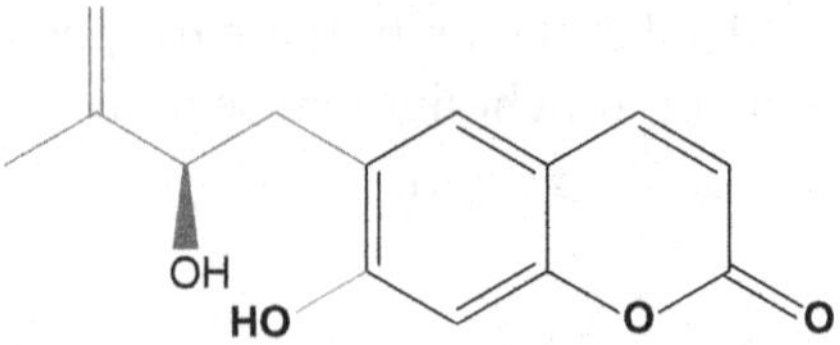

232

4.2.9 Functional Group at C$_7$ Position

4.2.9.1 Hydroxycoumarin (Umbelliferone)

Umbelliferone exhibited lower inhibitory effects for the b-secretase (BACE1).[60]

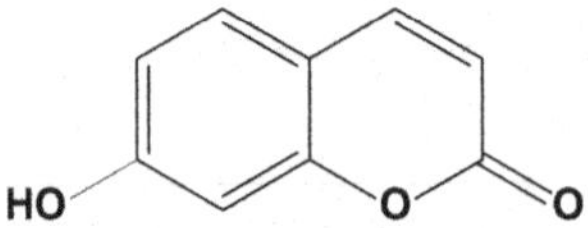

umbeliferone

4.2.9.2 Alkoxy Group

7-Methoxycoumarin [233] exhibited lower inhibitory effects for the b-secretase (BACE1).[60]

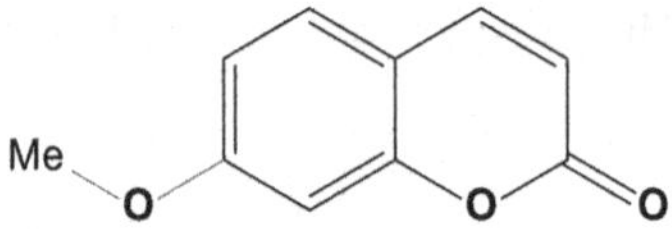

methylumbeliferone
233

Prenyloxy group

7-Prenyloxycoumarin [90] exhibited lower inhibitory effects for the b-secretase (BACE1).[60]

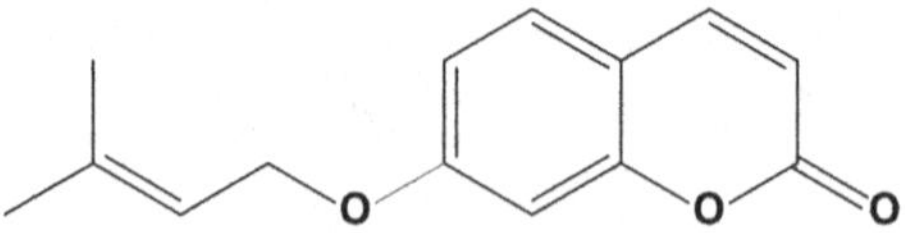

O-prenylumbeliferone
90

Geranyloxy group

The geranyloxy group at the C-7 position of simple coumarin [91] is very significant for activity. A geranyloxy group present at the C-7 position of aurapten showed moderate activity as the b-secretase (BACE1) inhibitor.[60]

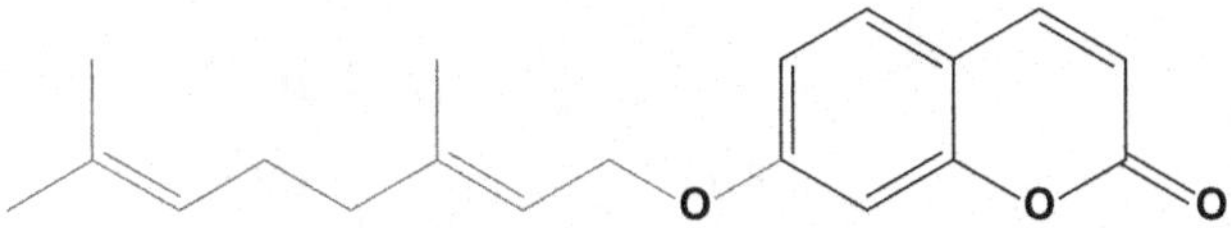

aurapten

91

4.2.10 Functional Group at C_6 and C_7 Position

In methoxy groups at C-6 positions [**217d**–**217e**] and at C-7 [**217b**–**217c**], for para and meta derivatives, the C-6 substituted derivatives appeared to have additional significance than the C-7 substituted derivatives in maintaining submicromolar activity. 6,7-dimethoxy coumarin and N-benzyl-N-methyl substituents in the para position deliberated the finest anticholinesterase activity.[63]

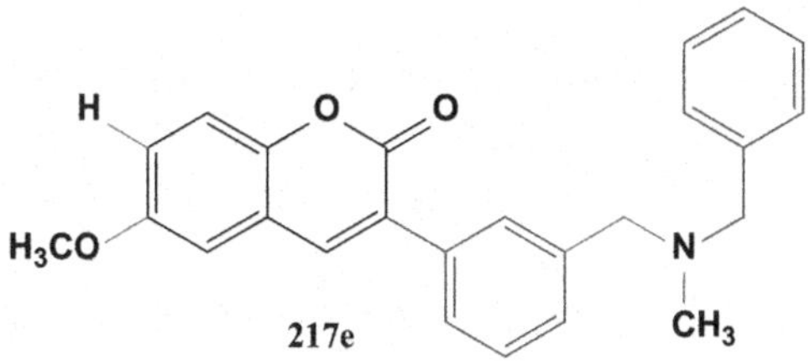

217e

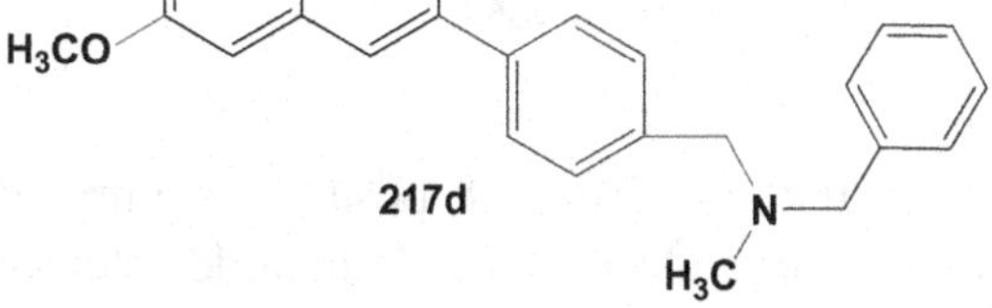

217d

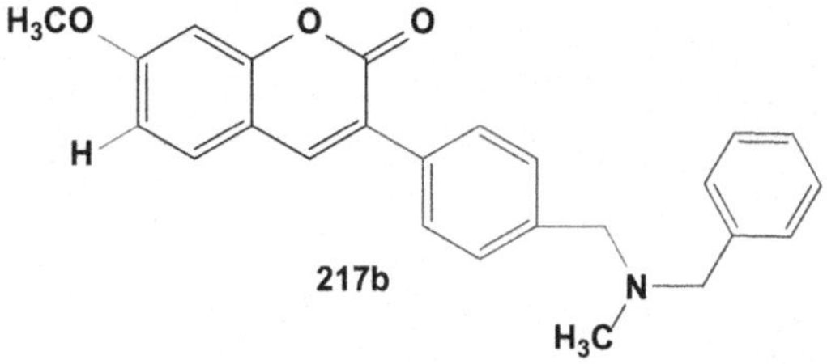

217b

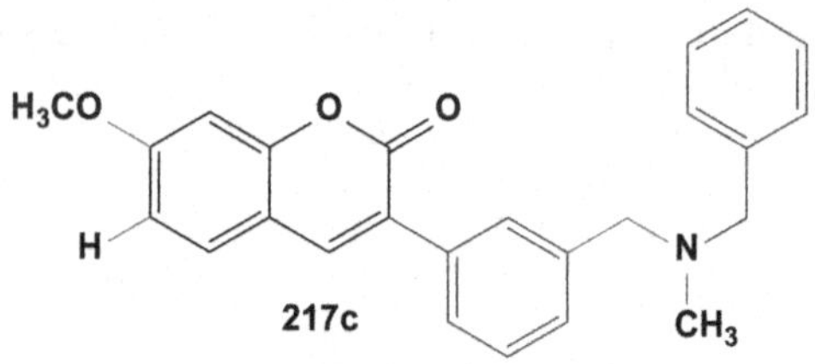

217c

4.2.11 Fused Ring at C$_6$-C$_7$ Position
4.2.11.1 Furanocoumarin

Psoralen [234], without the substitutions on the benzene ring, was established for a significantly lower effect.[60]

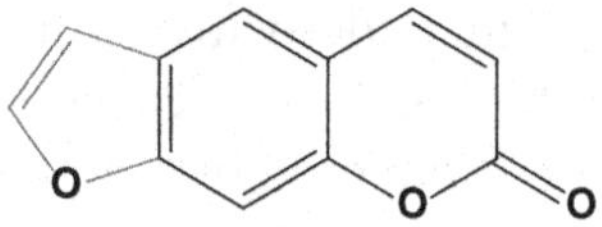

psoralen
234

Methoxy Groups

Two methoxy groups at the C-5 and C-8 positions of isoimpinellin [235] had no inhibitory activity.[60]

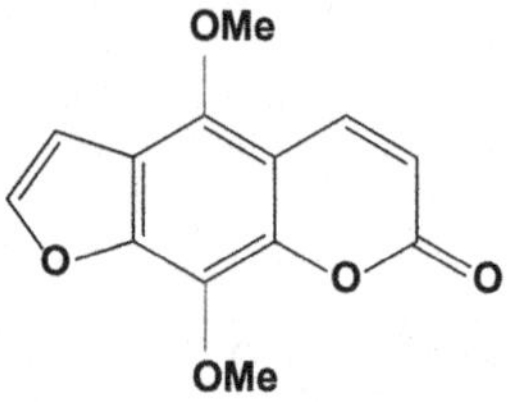

isopimpinellin
235

Prenyloxy Group

A prenyloxy group at the C-5 or C-8 position of imperatorin [236] displayed increased inhibitory effects, whereas an added methoxy group at the C-8 or C-5 position had no important influence as in phellopterin [237].[60]

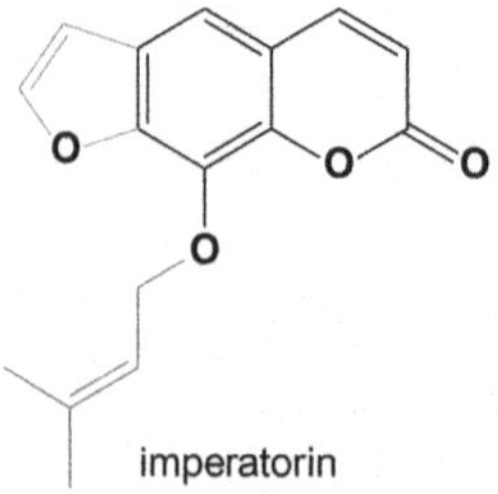

imperatorin
236

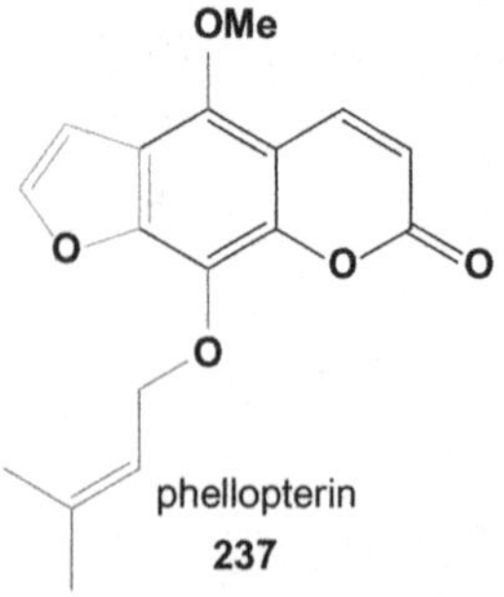

phellopterin
237

A prenyloxy group at the C-5 or C-8 position of isoimperatorin [191] and xanthotoxin [238] displayed increased inhibitory effects, whereas an added methoxy group at the C-8 or C-5 position had no important influence as in knidilin [239].[60]

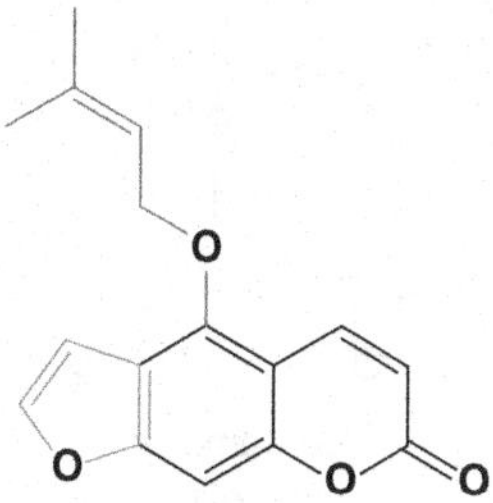

isoimperatorin
191

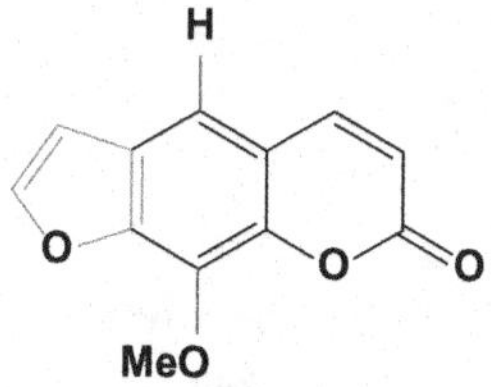

xanthotoxin
238

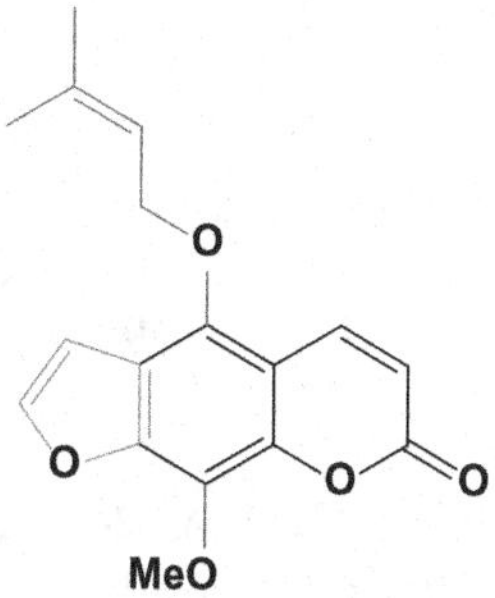

kinidilin
239

Geranyloxy Group

A geranyloxy group at the C-5 or C-8 position of 8-geranyloxypsoralen [240] and bergamottin [241] was established as more active than prenyloxy groups and an additional methoxy group at the C-8 or C-5 position.[60]

8-geranyloxypsoralen
240

bergamottin
241

Glucose Derivative

The glucose derivative had little effect in inhibiting AChE such as in nodakenin [**242**].[65]

242

4.2.11.2 Pyranocoumarins

Pyranocoumarins [**194, 195, 197, 244, 245, 247, 248**] did not exhibit any significant activity as BACE1 inhibitors.[60] Likewise, xanthyletin [**188**], pyranocoumarin with an absence of free hydroxyl moiety at position 3, was inactive; this elucidated that the presence of a free hydroxyl moiety at the C-3 position is significant in showing an inhibitory effect for AChE using pyranocoumarins. A dihydropyranocoumarin, decursinol [**243**] stayed a potent inhibitor of AChE. Decursin [**240**], with an oxygenated isoprenyl unit as a replacement for a free hydroxyl group at C-3 position, was a weaker inhibitor.[65]

194

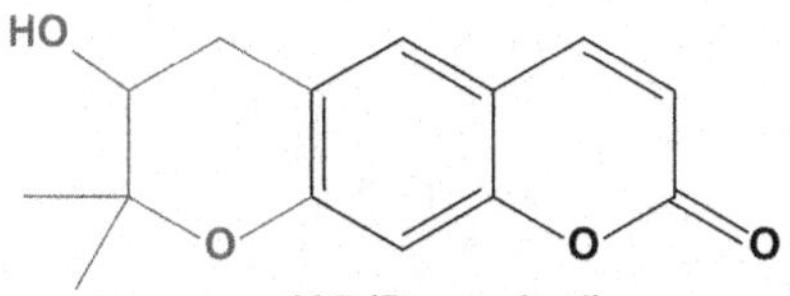

195 (Decursinol)

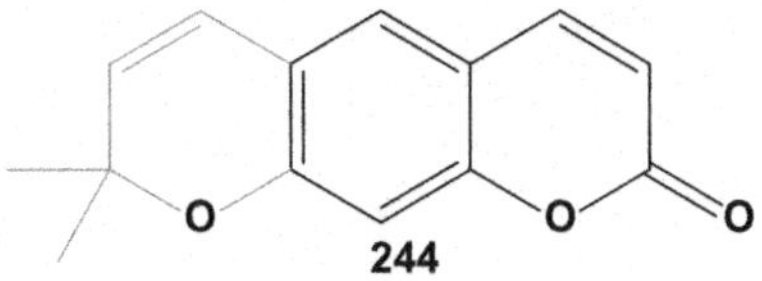

244

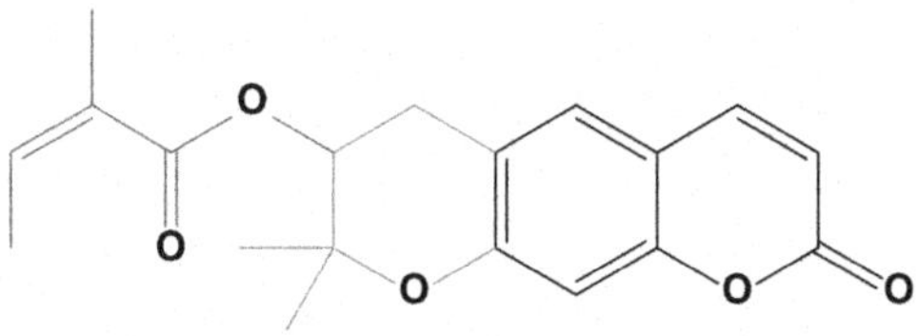

245

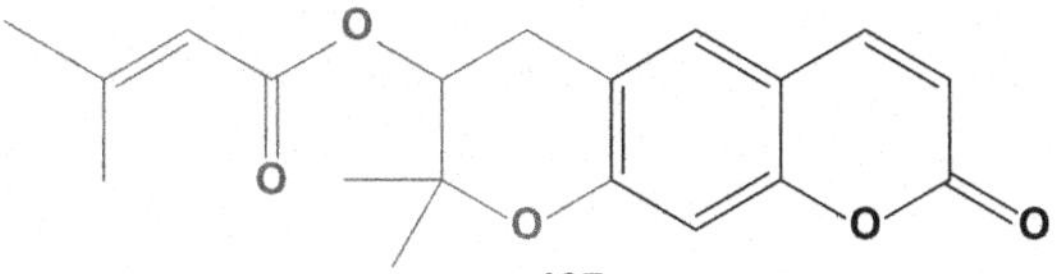

197

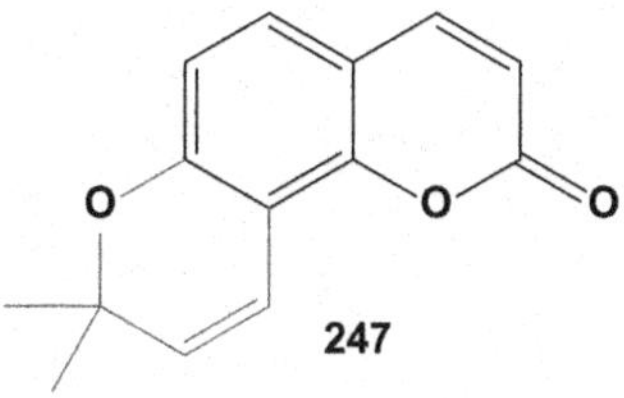

247

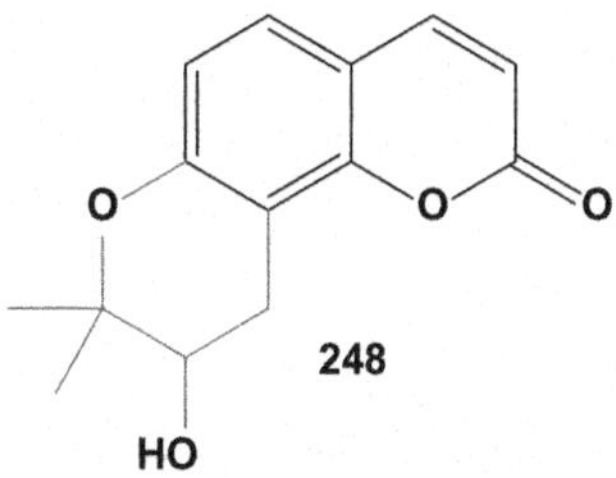

248

CHAPTER 5

Antidepressing Agents

Archi Sharma

Department of Chemistry, National Institute of Technology, Raipur, Chhattisgarh, India

5.1 INTRODUCTION

Figure 5.1 Antidepressants www.mentalhealthy.co.uk/news/277-research-into-the-effectiveness-of-using-two-anti-depressants-to-treat-depression.html.

Depression is a psychological disorder that brings depressed mood, feelings of regret, loss of pleasure or awareness, low self-confidence, disturbed sleep or appetite, low energy, and poor concentration. Agents that help to cure depression are known as *antidepressants* (Figure 5.1). Depression is not only a factor for diseases such as stroke and diabetes; it is also related with prolonged diseases as it proceeds to a depressed state, including discomfort, pain, diabetes, arthritis, heart diseases, cancer, and the like. Certain groups of chemicals in the brain called *neurotransmitters*, such as noradrenaline and serotonin, can improve mood and emotion. It is supposed that antidepressants work with the increasing level of these neurotransmitters, although this process is not fully understood. Increasing levels of neurotransmitters can also disrupt pain signals sent by nerves, which may explain why some antidepressants can help relieve long-term pain.

5.1.1 Functional Group at C₃ Position

5.1.1.1 Aryl Group

Meta and Para Substituents

Phenyl-substituted meta and para derivatives of coumarin [249–250] were in the most favorable positions for the preferred MAO-inhibitory effect.[66]

249

249	R₁	R₂	R₃	R₄
a.	H	H	H	H
b.	H	H	CH₃	H
c.	H	CH₃	H	H
d.	H	H	OCH₃	H
e.	H	OCH₃	H	H
f.	OCH₃	H	H	H
g.	H	OCH₃	H	OCH₃
h.	H	OCH₃	OCH₃	H
i.	H	OCH₃	OCH	OCH₃
J.	H	Br	OCH₃	H
k.	H	OCH₃	Br	H
l.	Br	OCH₃	H	OCH₃
m.	Br	OCH₃	OCH₃	OCH₃
n.	H	H	H	OH
o.	H	OH	H	H
p.	OH	H	H	H
q.	H	H	C₃H₅O₂	H

250

250	R₁	R₂	R₃	R₄	R₅
a.	OCH₃	H	CH₃	H	H
b.	OCH₃	H	H	CH₃	H
c.	OH	H	H	CH₃	H
d.	C₃H₅O₂	H	H	CH₃	H
e.	C₅H₉O	H	H	CH₃	H

5.1.2 Functional Group at C₄ Position

Monoamine oxidases are enzymes occurs in outer membrane of mito-chondria in glial and neuronal cells and catalyses the oxidative removal of amines from monoamine neurotransmitters.[67] This reaction occures specifically in those types of biogenic amines found in the brain and the peripheral tissues which regulates intracellular activity such as given in Figure 5.2. The MAO-A isoform has greater attraction for serotonin and noradrenaline, while hMAO-B isoform favorably deaminates benzyla-mine and β-phenylethylamines.[68]

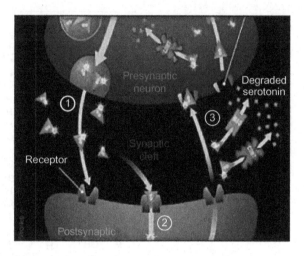

Figure 5.2 Function of Monoamine oxidase. http://depressivedisorder.blogspot.in/2010/02/monoamine-oxidase-inhibitors-maois.html.

5.1.2.1 Methyl Group

The presence of mono substitution in the C_4 position with a CH_3 group was directed to more active derivatives [251a−b], whereas phenyl [251c], trifluoromethyl [251d], and hydroxyl [251e] groups at the same position have decreased activity.[69]

251	R_1	R_2	R_3	R_4
a.	H	CH_3	H	$OCH_2C_6H_5$
b.	H	CH_3	H	$OCH_2C_6H_4\text{-}3'\text{-}NO_2$
c.	H	C_6H_5	H	$OCH_2C_6H_5$
d.	H	CF_3	H	$OCH_2C_6H_5$
e.	H	OH	H	$OCH_2C_6H_5$

5.1.2.2 Hydroxyl Group

The presence of extremely polar anionic groups such as hydroxyl groups at C-4, along with the linear alkoxy substituents, gave a set of effective and selective MAO-B inhibitors. 4-hydroxy coumarin derivatives, which are fully ionized at physiological pH, revealed low MAOB inhibitory activity. Examples of such derivatives are the compound without any substitution at 3′postion [252] and the 3′-fluoro-benzyloxy derivative [253].[70]

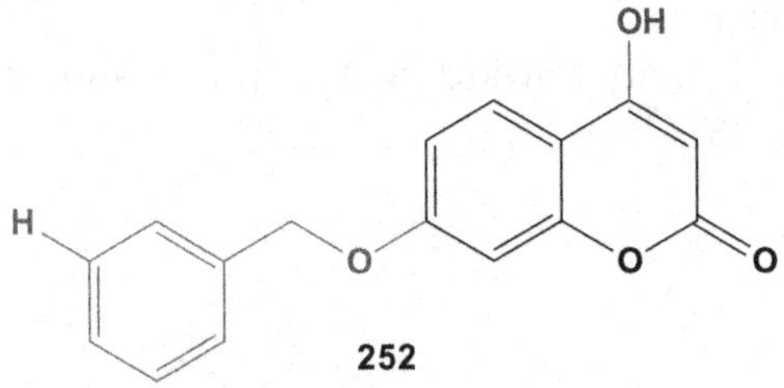

252

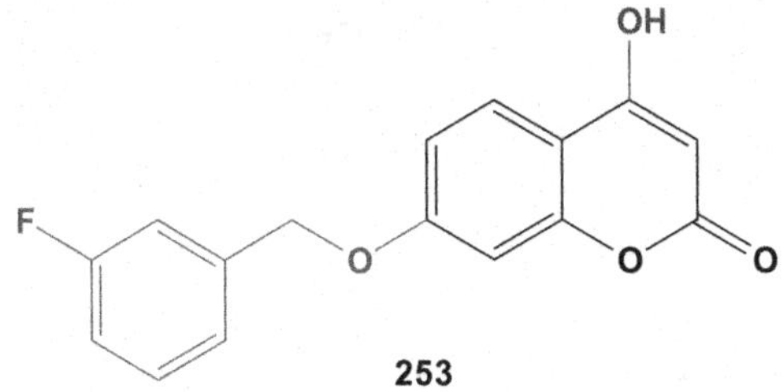

253

5.1.2.3 Carboxylic Group

Anionic carboxylic acid derivatives [**254–255**] displayed almost no MAO inhibition.[70]

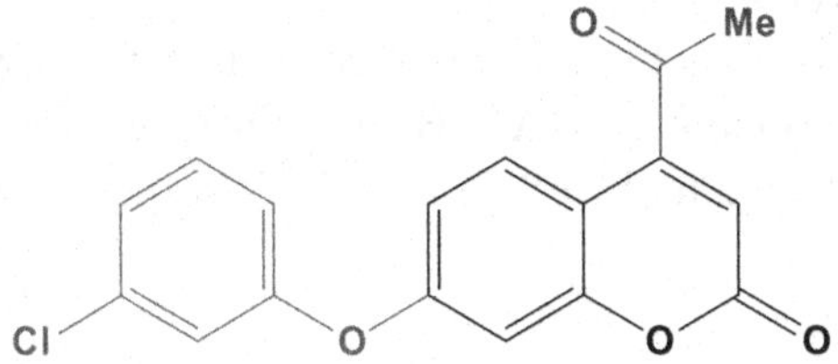

254

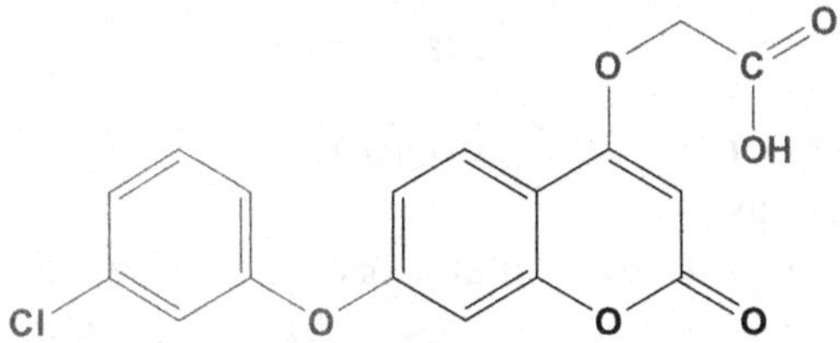

255

5.1.2.4 Ester Group

Some inhibitory activity toward MAO-B was shown by ethyl ester derivatives [256–257].[70]

256

257

5.1.2.5 Amide Group

4-oxyacetamido-7-metachlorobenzyloxy coumarin [258] is most effective as well as selective MAO-B inhibitor, exhibiting outstanding pIC50 and SI values equal to 8.48 and 3.86, respectively.[70]

258

5.1.3 Functional Group at C_6 Position

5.1.3.1 Small Groups

6-substituted-3-(pyrrolidin-1-ylmethyl)coumarins were determined for human monoamine oxidase A (MAO A) and B (MAO B) inhibitory activity. The most potent and selective coumarins are the 6-subsituted 3-(pyrrolidin-1-ylmethyl)chromen-2-ones, with small substituents such as OH [259], OMe and OCF_3 in the C-6 position. Among them,

6-hydroxy compounds were the most potent inhibitors of MAO-A (IC50 1.46 mM).[71]

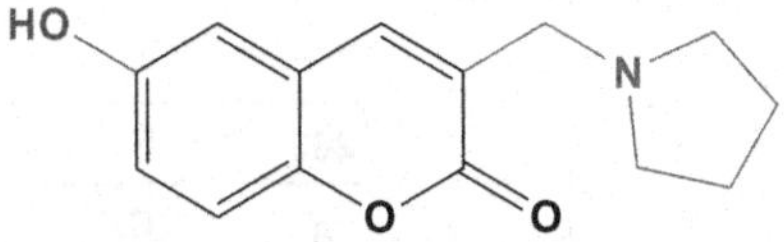

259

6-amino-substituted-3-(pyrrolidin-1-ylmethyl)chromen-2-ones derivatives [260] were also the most effective inhibitors of MAO-A (IC50 3.77).[71]

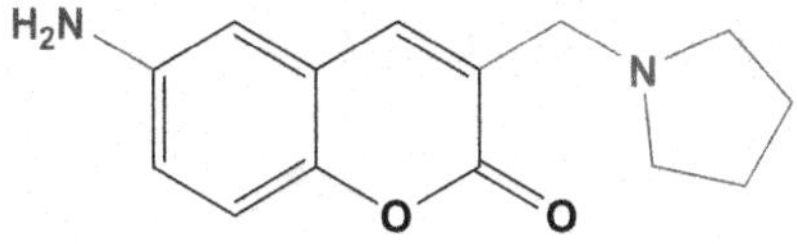

260

5.1.3.2 Some Large Group
Larger groups at the C-6 position drastically decrease the MAO-B inhibitory activity; thus compounds [261−262] have reduced efficiency.[66]

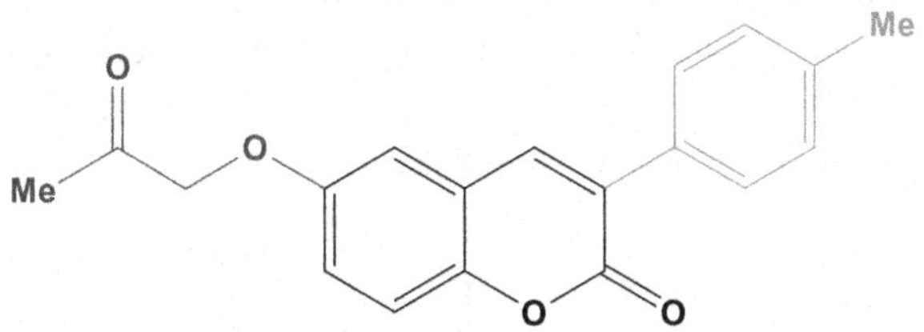

261

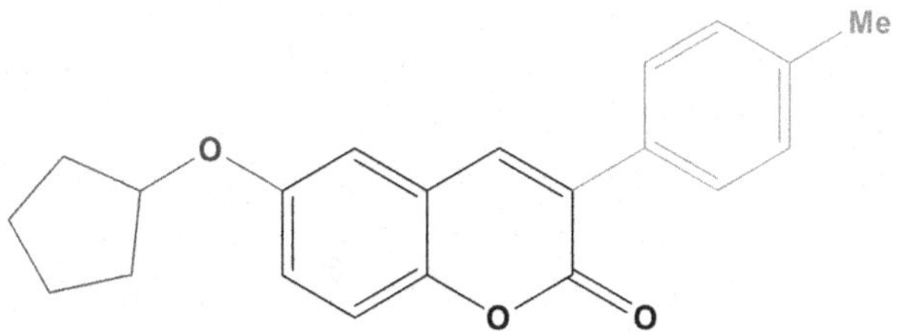

262

Similarly, it is clear that the presence of substituents in the C-6 position, irrespective of the size and lipophilicity, eradicated the inhibitory effect, as in compounds [263a–d].[69]

263	R_1	R_2
a.	OCH_3	$OCH_2C_6H_5$
b.	$OCH_2C_6H_5$	OH
c.	glucosyl	OH
d.	glucosyl	$OCH_2C_6H_5$

263

Benzyloxy Group
The corresponding 6-benzyloxy 3-(pyrrolidin-1-ylmethyl)chromen-2-ones [264] were found to be inactive.[71]

264

5.1.4 Functional Group at C_7 Position
5.1.4.1 Benzyloxy Group
Benzyloxy substituents at the 7-position [265–266] caused perfect interactions within the MAO-B binding pocket. Inversion of the direction of the OCH_2 functionality led to a slight decrease in MAO-B and a significant increase in MAO-A inhibitory activity.[69]

265

266

Length of Ether Bridge

A comparison with ether-connected chains of isomeric coumarin pairs showed the 6-isomers to be more selective for MAO-B over MAO-A, whereas 7-isomers tended to be more potent at MAO-B.[72] Increasing the length of the ether bridge did not significantly change activity. The comparison of the activity of derivatives and esculin clearly reveals the effect of steric hindrance at position 6; hence only small substituents such as OH are tolerated.[69]

267	R_1	R_2
a.	OCH_3	$OCH_2C_6H_5$
b.	$OCH_2C_6H_5$	OH
c.	Glucosyl	OH

268	R_1	R_2
a.	OH	$OCH_2C_6H_5$
b.	$OCH_2C_6H_5$	OH
c.	$OCH_2C_6H_5$	$OCH_2C_6H_5$

Electron-withdrawing Group

Compounds with strong electron-withdrawing groups [269a–h], such as NO_2, CN, regardless of their position, showed MAO-A inhibitory activity higher than lead [270]. Halogenated compounds showed effective inhibitory activity.[69]

269	X
a.	2-CN
b.	$3,5\text{-}(NO_2)_2$
c.	3-F
d.	3-Cl
e.	$3\text{-}NO_2$
f.	3-CN
g.	$4\text{-}NO_2$
h.	4-CN

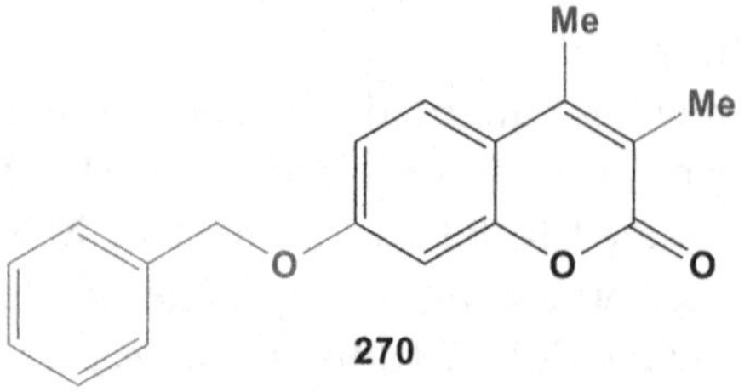

270

Ortho Substituent

Ortho substitution looked unfavorable, on comparison between possible positional isomers of CH_3 and CN substituted derivatives.[69]

Para Substituents

Halogen at para substitution [271i–l] showed the highest MAO-A inhibitory activity.[69]

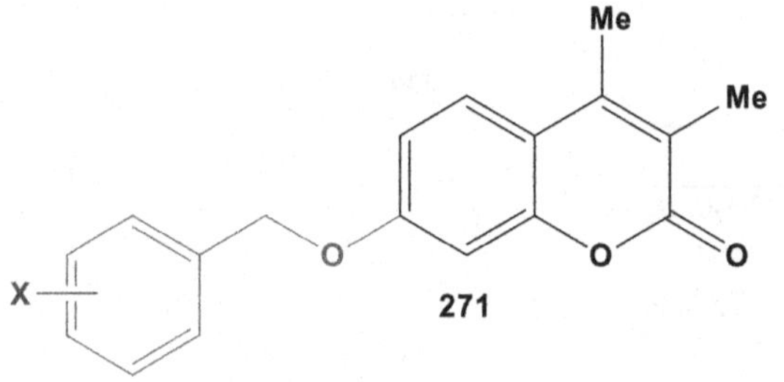

271

271	X
i.	4-F
j.	4-Cl
k.	3,4-F2
l.	3-OH,4-F

Halo-benzyloxy Group

Submicromolar inhibitory activity at MAO-B was displayed by the 3′-chlorobenzyloxy coumarin [272] and 3′-bromobenzyloxy coumarin [273].[70]

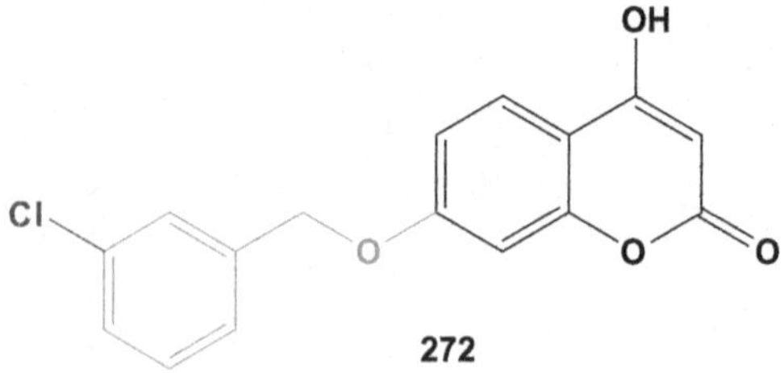

272

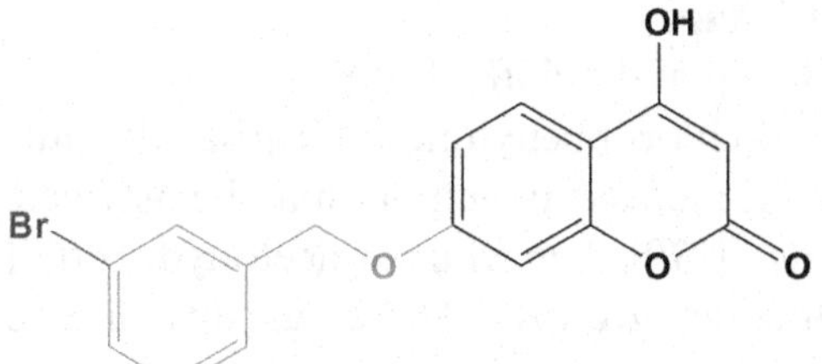

273

Bulkier Group

It is also important to note that the benzyloxy derivatives bearing bulkier substituents at the C-7 position [274–275] decrease the activity in MAO-A inhibition due to steric hindrance.[69]

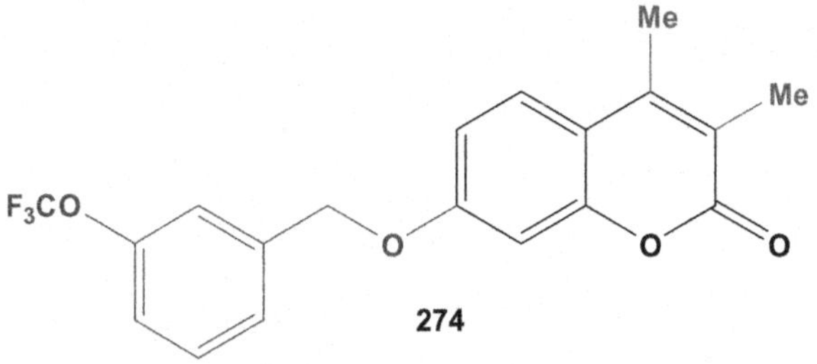

274

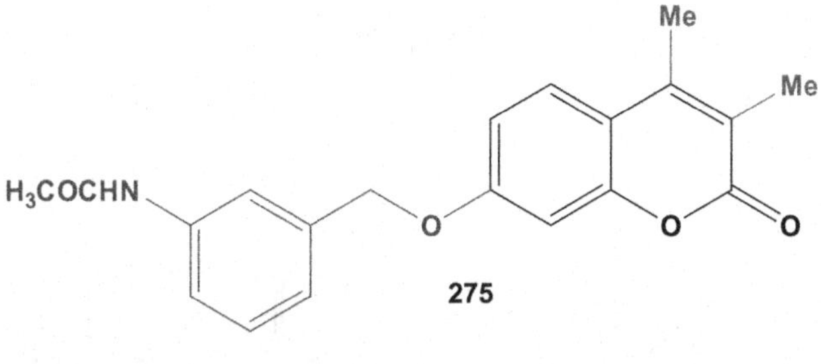

275

5.1.4.2 Hex-5-ynyloxy Group

At position 7 a hex-5-ynyloxy chain [276] was suitable to be given strong MAO-B inhibition.[72]

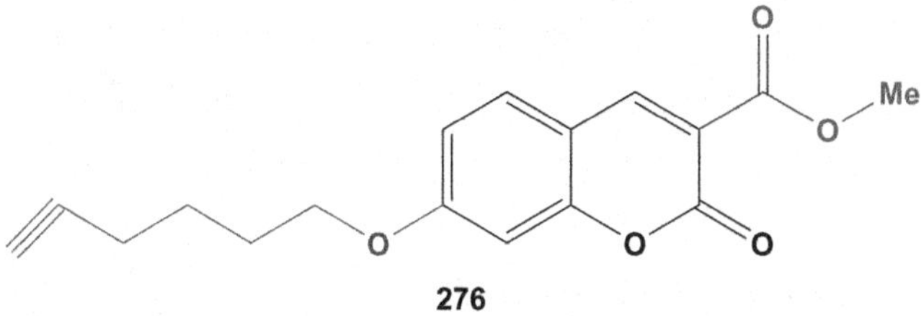

276

5.1.4.3 Alkoxy Linkage
Chlorophenyl Group and Schiff's Base
Introduction of 4-chlorophenylthiosemicarbazides and Schiff's bases showed that a mild increase in activity may be ascribed to their flexible thiosemicarbazides [279], benzylidene acetohydrazide [278], or ethylidene acetohydrazide groups. These groups are appropriate for interaction in hydrogen binding, whereas increase in activity of 4-(4-chlorophenyl)-5-mercapto-1,2,4-triazoles [277] was observed which may be due to aryl fragment leads to hydrophobic interactions and electrostatic surface attraction with the active sites of MAO isoenzymes under the hydrophobic region.[73]

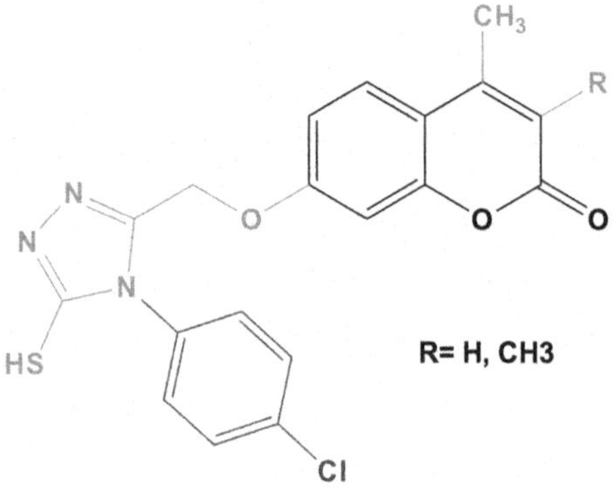

R= H, CH3

277

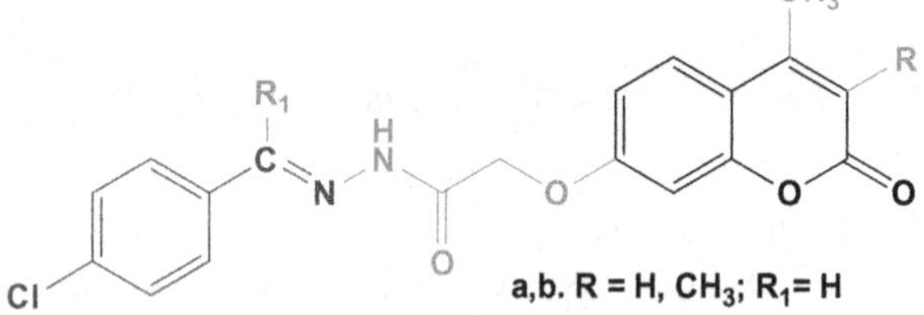

a,b. R = H, CH3; R1= H
c,d. R = H, CH3; R1= CH3

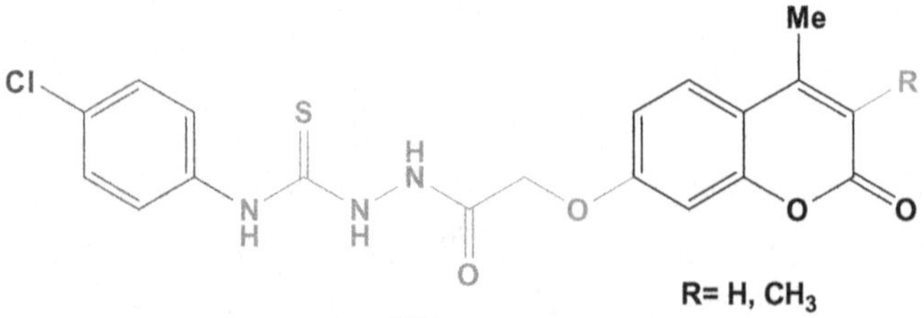

278

R= H, CH3

279

Oxadiazole and Thiadiazole Group

A mild inhibitory activity for MAO-A and MAO-B was shown by the mercapto-1,3,4-oxadiazole [280] and their thiadiazole bioisosters [281−282]. This was expected due to the inserted aryl fragment accommodated under the hydrophobic region of MAO isoenzymes and showed hydrophobic interactions and electrostatic surface attraction with the respective active sites.[73]

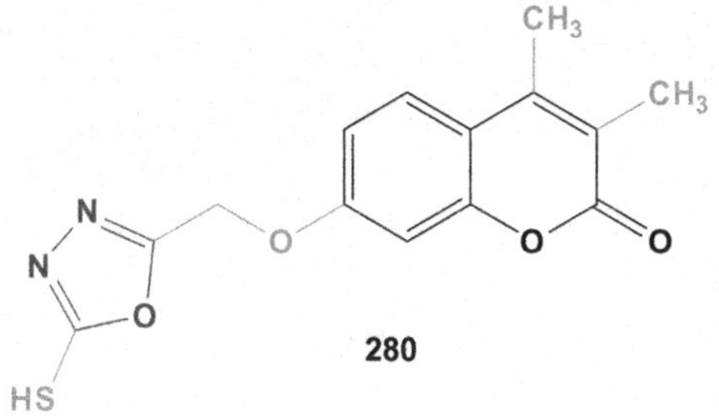

280

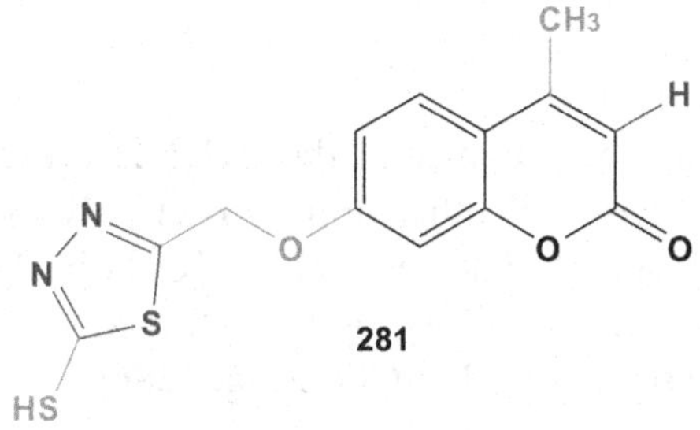

281

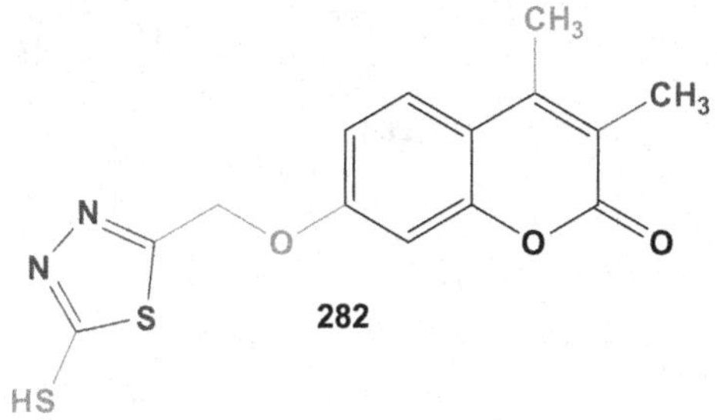

282

Thiazolidinone Group

The isosteric variations of 2-(4-chlorophenylimino)-4-oxothiazolidine derivatives [283] have a small enhanced affinity selective for MAO-B rather than MAO-A. Isosteric changes at the C-7 position of the considered coumarin derivatives bring back suitable MAO inhibitory affinity as well as selectivity between the MAO-A and MAO-B.[73]

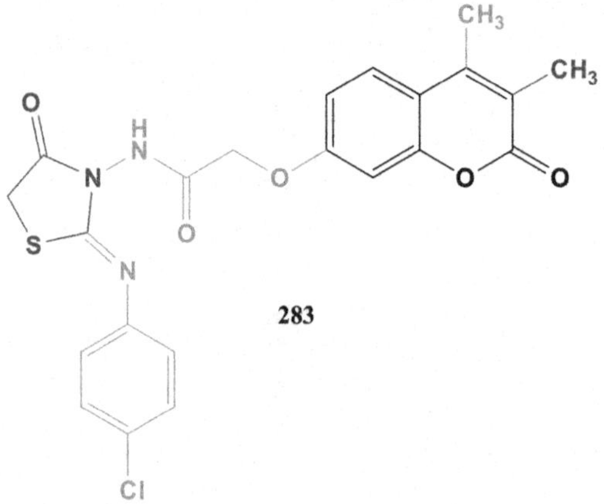

283

5.1.4.4 NHCH₂ Linkage

At the C-7 position, the presence of the CH_2NH bridge [284] decreased both MAO-A and MAO-B inhibition, particularly much lower inhibitory activity for MAO-B. The reversed $NHCH_2$ bridge [285] also produced similar effects. Inhibitory activity for MAO was retained on replacement of $NHCH_2$ to a NHCO bridge [286].[69]

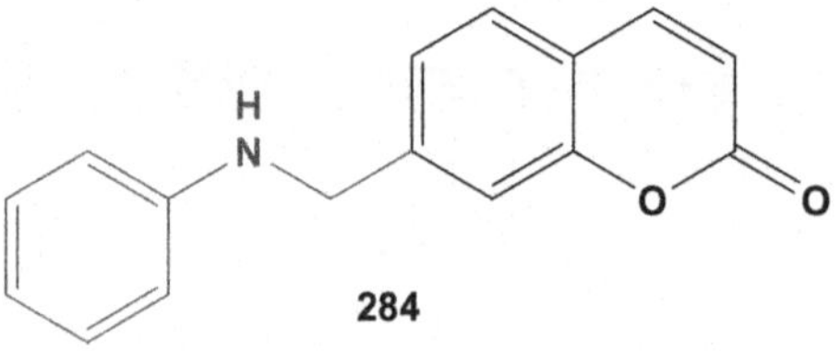

284

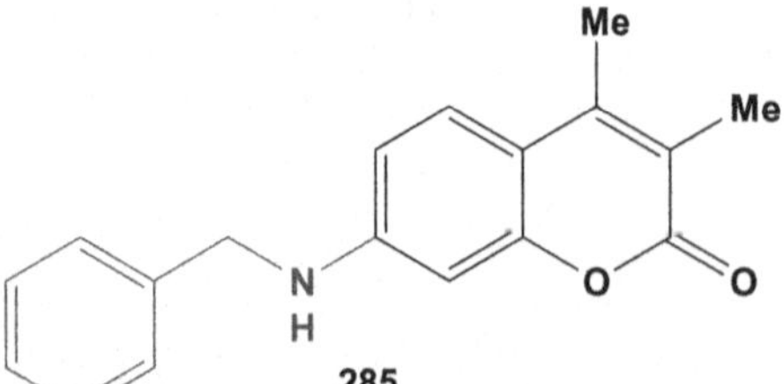

285

286

5.1.4.5 NHSO$_2$ Linkage

The presence of isosteric NHSO$_2$ groups as such in compound [287] lost both MAO inhibitory activities, which is expected due to the ionizable group.[69]

287

5.1.4.6 O-SO$_2$ Linkage

The selectivity for two enzymes in coumarin inhibitors was inverted by an O-SO$_2$ group. MAO-B inhibitory activity of [288b–e] is higher than MAO-A inhibitory activity compared with the [288a].[69]

288

288	R
a.	$OCH_2C_6H_5$
b.	$OSO_2C_6H_5$
c.	$OSO_2C_6H_4$-4'-CH$_3$
d.	$OSO_2C_6H_4$-4'-OCH$_3$
e.	$OSO_2C_6H_4$-4'-NO$_2$

5.1.4.7 Alkenoxy Linkage

Removal of the methyl group, as in compound [289] on the alkenoxy bridged derivative, showed the highest MAO-B inhibitory.[74]

289

5.1.5 Disubstituted or Fused Functional Group
5.1.5.1 Disubstituted Group at C₃ and C₄ Position

Substitution at C-3 and/or C-4 positions of the coumarin ring affects inhibitory activity with selectivity for the two-enzyme MAO-A and MAO-B.

Methyl Group
The presence of two methyl groups at C-3 and C-4 positions of the coumarin nucleus [290] gave better inhibitor activity.[69,74]

270

3,4-dimethylgeiparvarin [290] does not exhibit any improved MAO-B inhibitory effect. This is due to different binding sites of the two MAO-A and B inhibitors because of the presence of longer, more flexible, and bulkier substituents at position 7 of coumarin. Greater MAO-B inhibitory effect of geiparvarin is due to the occurrence of a methyl group on the double bond, perhaps cause for undesirable steric outcomes prominent to lesser affinity.[74]

3,4-dimethylgeiparvarin
290

Pheny and Methyl Group

In the coumarin derivative with a phenyl group at the C-3 position and a CH_3 group at the C-4 position [291], the absence of both MAO-A and B inhibitory effect was found. MAO-B binding sites have to receive small lipophilic groups at C-3 and C-4 positions of the coumarin nucleus whereas larger hydrophobic (CF_3, phenyl) or hydrophilic (OH) are not well tolerated. In the case of CF_3 group other electronic factors may interfere.[69]

291

Hydrogenation

Hydrogenation of the 3,4-double bond as such in compound [292] abolished the MAO inhibitory activity.[69]

292

5.1.5.2 Annelation at C_3 and C_4 Position

3,4-Annelation using five- and six-membered cyclic structure was acceptable only for cyclo-pentenyl [293] and alicyclic [294] rings.[69]

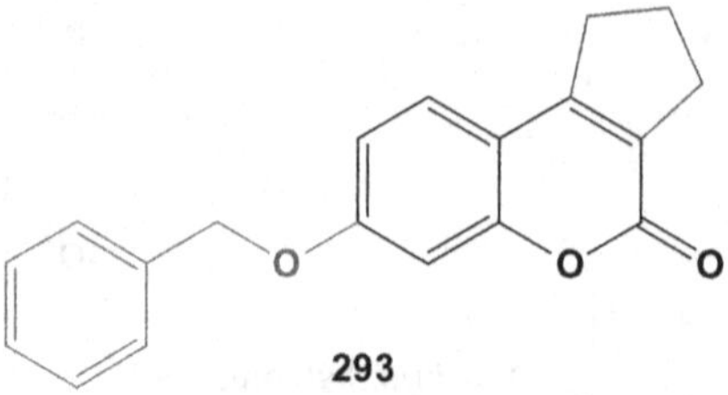

293

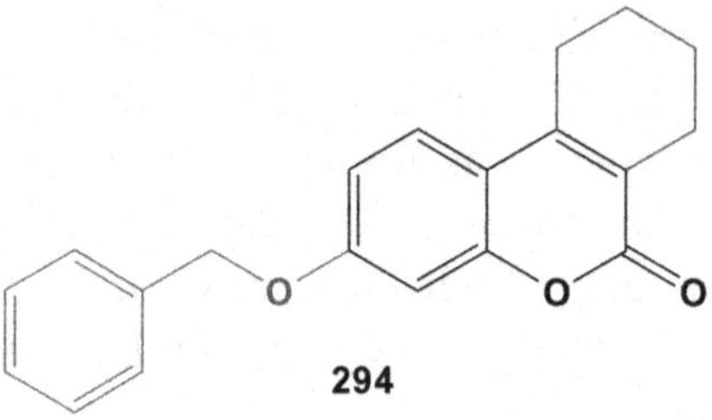

294

5.1.5.3 Substitutes at C_3 and C_6 Position

3-Arylcoumarins are potent and selective monoamine oxidase B inhibitors. It was found that the presence of phenyl groups at C_3 of the coumarin nucleus, with a small substitution at C_6 (methyl or methoxy groups), as in compounds [249, 250a−b], seems to be important.[66]

CHAPTER 6

Anti-Inflammatory Agents

Archi Sharma

Department of Chemistry, National Institute of Technology, Raipur, Chhattisgarh, India

6.1 INTRODUCTION

Inflammation, illustrated in Figure 6.1, is part of immune response of the body we can't heal without it. It is common in all types of diseases. The most common symptoms of acute inflammation are swelling, heat, pain, redness, and loss of functionality as illustrated in Figure 6.1. But when inflammation is out of control or prolonged, then it can lead to the several chronic diseases such as cancer, malaria, typhoid, diabetes, and many more given in Figure 6.2. Inflammation is part of the complex biological reactions of vascular tissues against harmful stimuli, such as damaged cells, pathogens, or irritants. The inflammatory responses are activated when damaged or infected cells release a

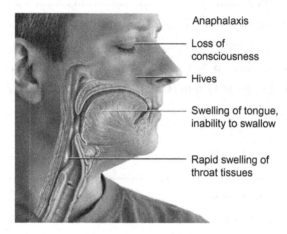

Figure 6.1 Inflammation. www.sccgov.org/sites/ems/Clinical%20Care/Documents/4-24-08Anaphylaxis.pdf.

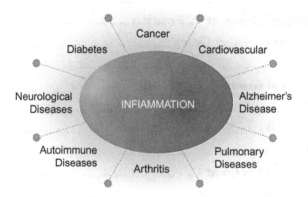

Figure 6.2 Chronic Inflammation. http://alaskahealth.info/inflammation/

chemical signal in the bloodstream. Anti-inflammatory drugs are most commonly taken by patients to reduce the inflammation in the body by eradicating the harmful stimuli as well as to initiate the healing process. Anti-inflammatory medicines include about half of analgesic drugs, relieving pain by decreasing inflammation that affects the central nervous system.

6.1.1 Functional Group at C_3 Position

6.1.1.1 Alkyl Group

Coumarin-based carbamates displayed potent inhibitory effects against production of TNF-α, stimulated by bacterial lipopolysaccharide (LPS) in human peripheral blood mononuclear cells (PBMC cells). It was found that a methyl group [295] or a proton [296] at carbon-3 of coumarin were not effective for any TNF-α inhibitory effect. This activity increases with increase in the size of the alkyl group, particularly with the size of carbocyclic moiety [297].[75]

295

296

297

6.1.1.2 Aryl Group

The TNF-α inhibitory effect significantly improved when the aryl group substituted at position 3 by methyl linkage [298]. Elimination [299] or addition [300] of this carbon chain to methyl group both eradicated the inhibitory activity.[75]

298

299

300

Tumor necrosis factor alpha (TNF-α), also known as TNFSF2 and cachec-tin, showed a crucial role in inflammation, immune system development, and apoptosis. TNF is a type of cytokine derived from monocyte. It is a significant intermediary between infection and the body's response. It stimulates inflammation as well as the destruction of the tissue in immune-facilitated diseases. TNF-α is also used in treatment of diseases, such as spondylitis, rheumatoid arthritis, psoriatic arthritis, Crohn's disease, and juvenile rheumatoid arthritis. The TNF-α protein is produced as a prohormone with a remarkable long and different signal system, which is not present in the mature cytokine. Small hydrophobic expanse of amino acids helps release the TNF in lipid bilayers.

6.1.1.3 Benzyl Group with Different Substitution

Halide Group

Inhibition of the TNF-α factor was found to be increased when a fluorine atom was present at the para position of the substituted C-3 benzyl group [293]. However, an ester group, carboxylic acid, or trifluoromethyl group was not effective at this position.[75]

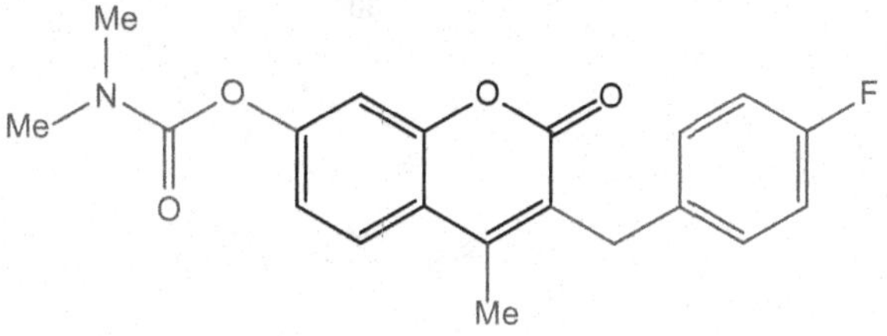

301

Acetamido and Ethyl Uriedo

An acetamido [302] or an ethyl ureido [303] group present at the meta position of a substituted C-3 benzyl derivative was as potent as the lead compound [298].[75]

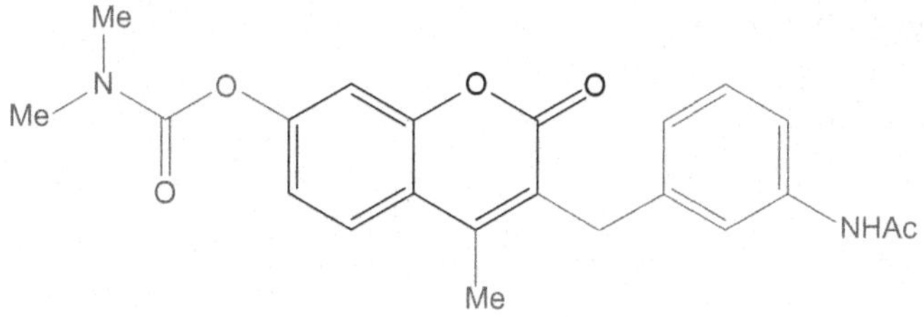

302

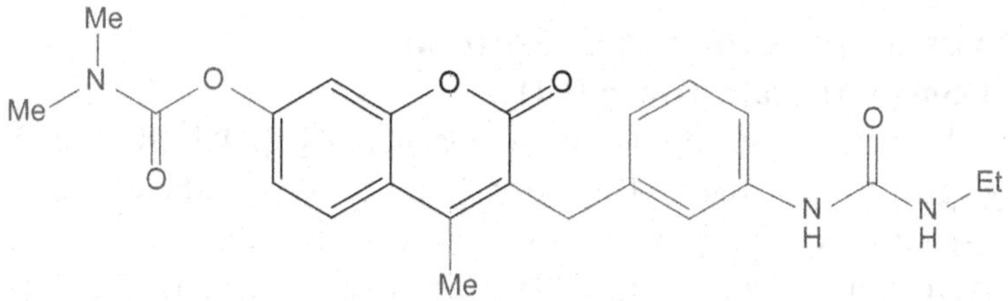

303

6.1.1.4 Heterocyclic Group

Heterocyclic analogs such as 1-piperidinylmethyl [304] or 4-morpholinylmethyl [305] substituents were not accepted at position 3.[75]

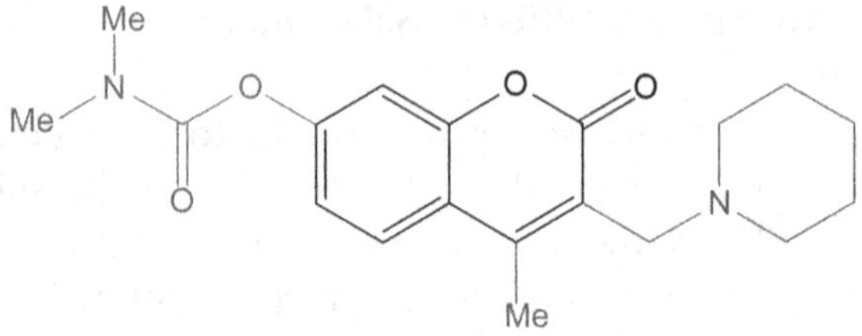

304

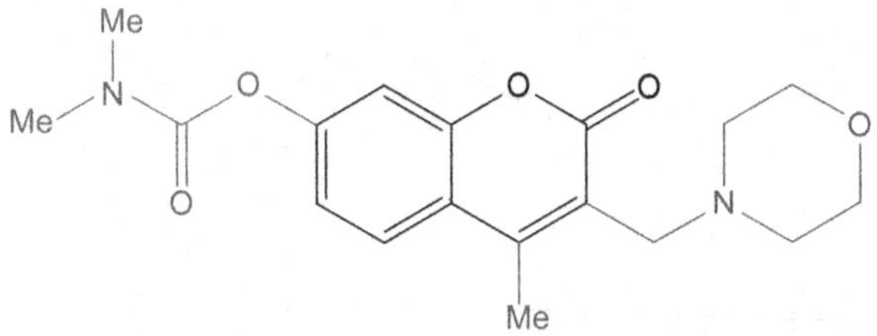

305

In *in vitro* assays furyl derivative [306] showed excellent inhibitory effect to avoid the formation of leukotrienes over the human 5-lipoxygenase enzyme.[76]

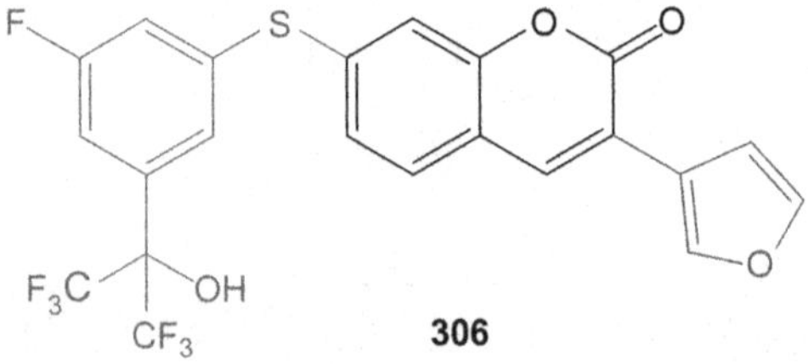

306

6.1.2 Functional Group at C_4 Position

6.1.2.1 Small Aliphatic Group (CH_3)

A small aliphatic group is finest for activity [298]. Removal of a methyl group from the C-4 position [307] totally eradicated the activity. *N*-propyl or hydroxyl ethyl substitutes exhibited comparable activity to that of compound [298]. Removal of a methyl group from the C-4 position and addition of trifluoromethyl or carboxylic acid resulted in inactive compounds.[75]

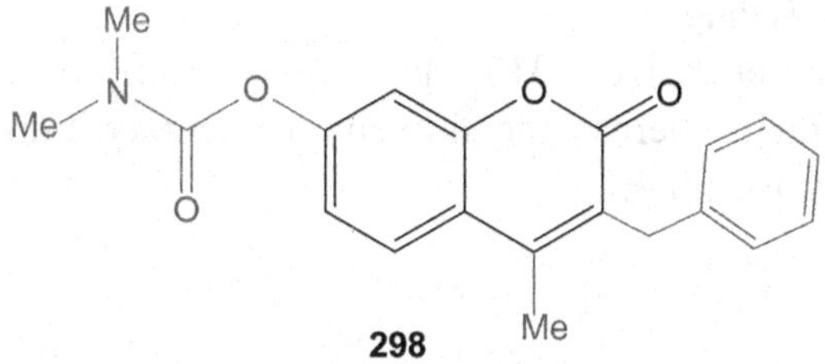

298

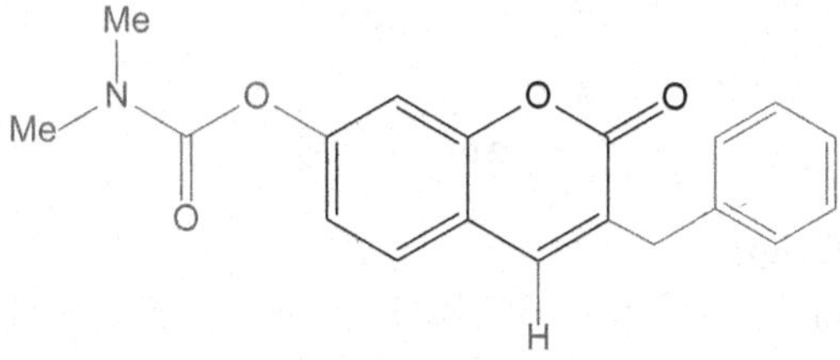

307

6.1.2.2 Hydroxyl Group

Complete loss or reduction in TNF-α inhibitory activity was shown by those analogs with a linear aliphatic group. A hydroxyethyl group such as in compound [308–309] was also found to be acceptable at position 4 to generate a potent TNF-α inhibitor.[75]

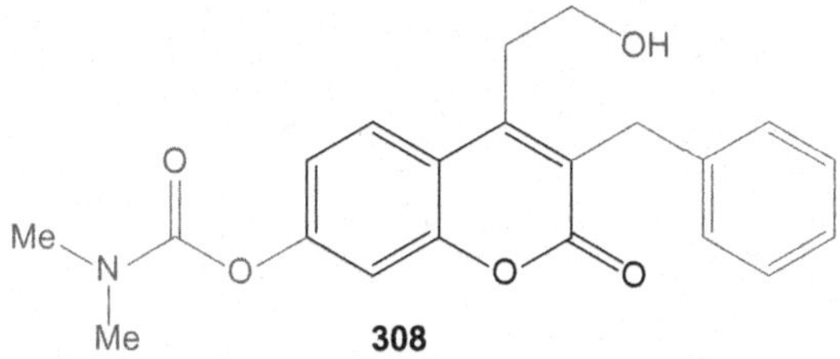

308

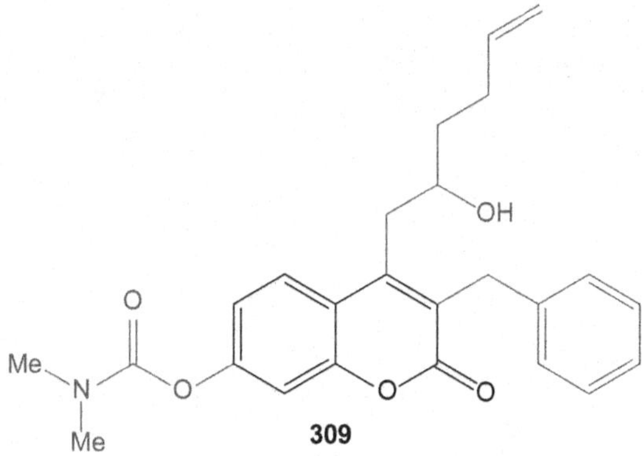

309

6.1.2.3 Alkoxy Group

The coumarin derivatives [310], lacking a furan ring, showed IC50 values of 890 nM, whereas the side-chain phenoxyl-butoxy group was substituted at position 4.[77]

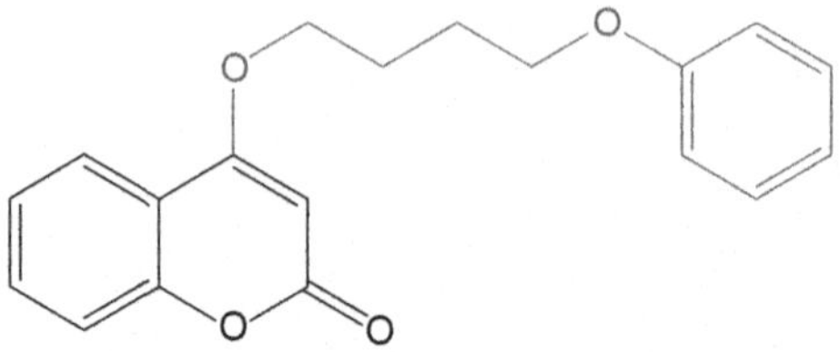

310

6.1.2.4 Cyclic Group

Thorough deficit or decrease of the TNF-α inhibitory effect was shown by those analogs having cyclohexyl [311], phenyl [312], substituted phenyl [313], heterocyclic [314], or linear aliphatic group.[75]

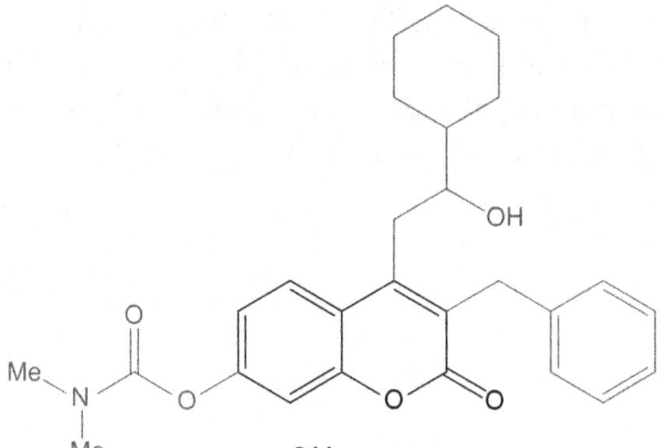

311

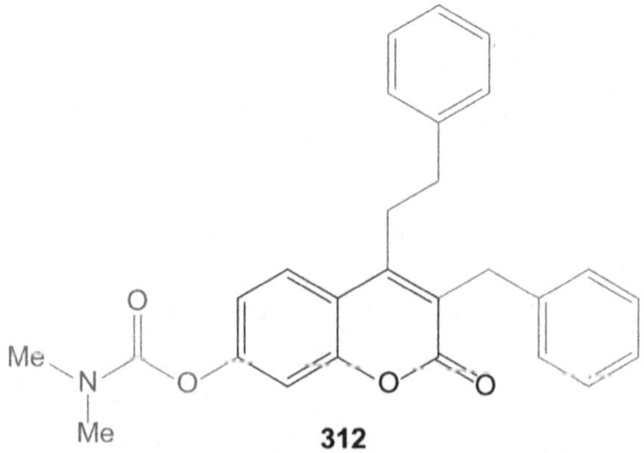

312

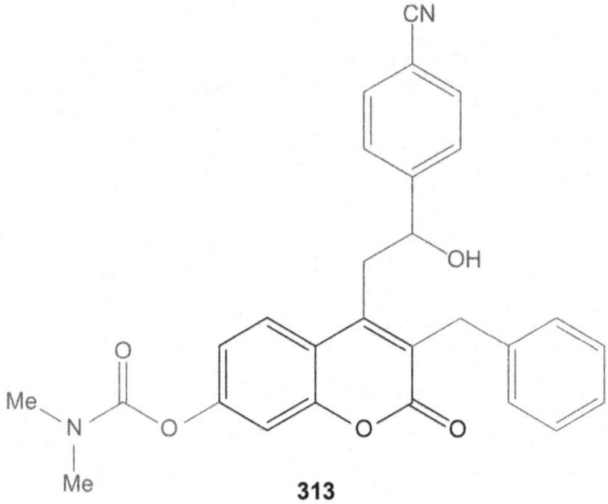

313

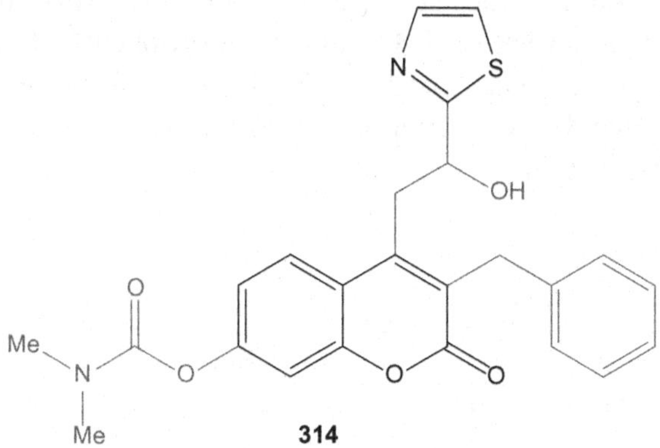

314

6.1.2.5 Aryl Group

The activity decreased when substituents such as fluorine atoms [315] or trifluoromethyl groups [316] were substituted at the para position of the phenyl group. Likewise, replacement of a methyl or ethyl group by pyridinyl groups was not effective for inhibitory activity.[75]

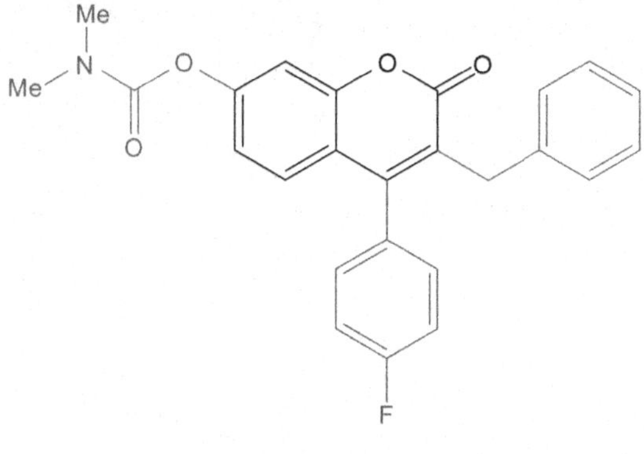

315

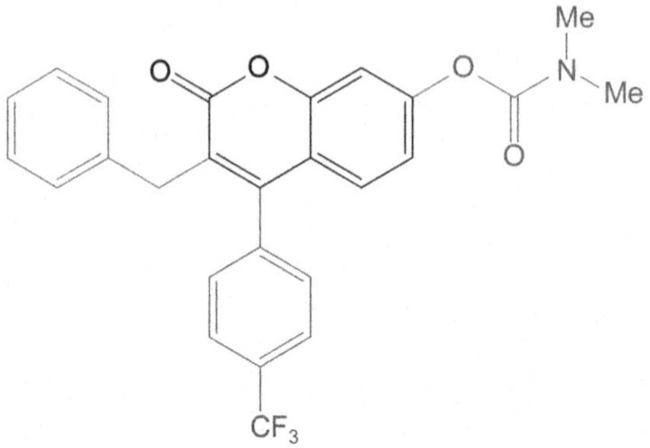

316

Except for the *p*-chloro-substituted compound [**319**], all derivatives such as phenyl derivative [**317**] and *p*-fluorophenyl derivative [**318**] *in vitro* assays revealed excellent inhibitory effects regarding their capacity to avoid the formation of leukotrienes over the human 5-lipoxygenase enzyme.[76]

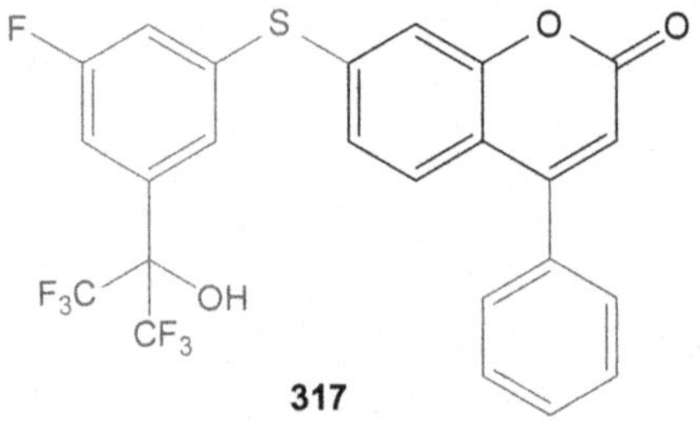

317

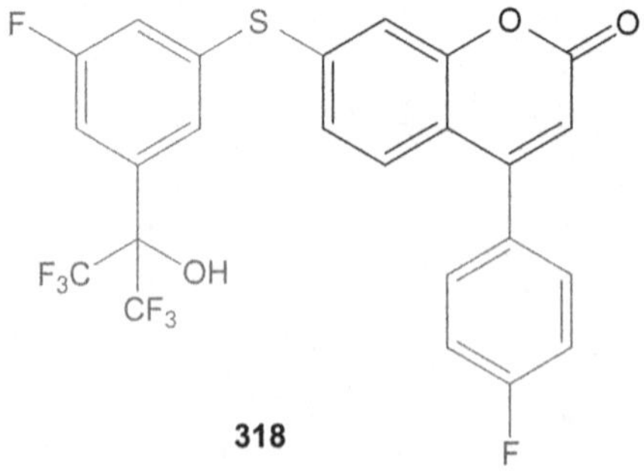

318

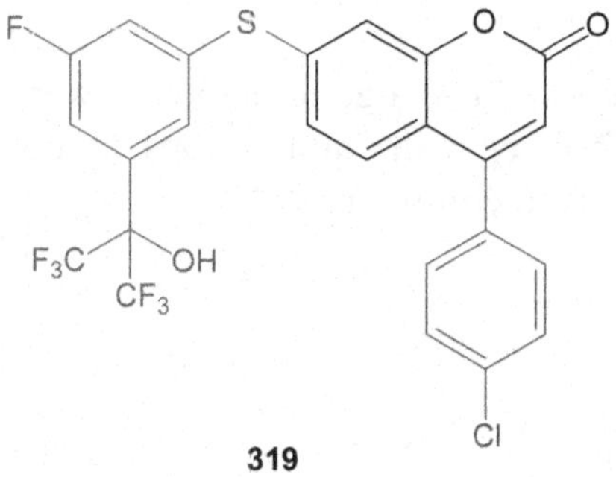

319

6.1.2.6 Pyrimidine Group

Pyrimidine derivatives substituted with −SH, -OH, and -NH$_2$ moieties were discovered for analgesic and antipyretic activities.[78]

Hydroxyl Group
Compounds [320a−e] containing -OH substituents at position 2 of the pyrimidine nucleus improve cleavage of the DNA.[78]

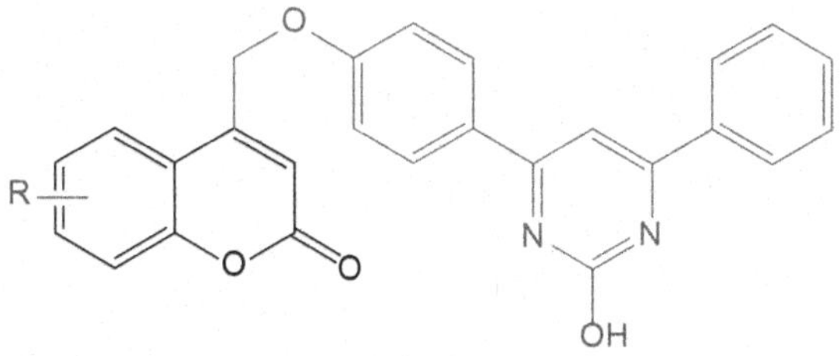

320	R
a.	6-CH$_3$
b.	7-CH$_3$
c.	6-Cl
d.	5,6-Benzo
e.	7,8-Benzo

320

Amine Group
The 2-position of pyrimidine moiety substituted with a -NH$_2$ group [321a−e] presented important antipyretic and analgesic activities related to the -OH and -SH groups.[78]

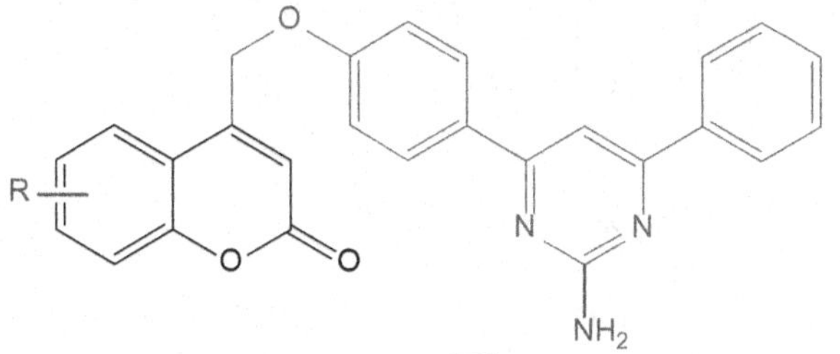

321	R
a.	6-CH$_3$
b.	7-CH$_3$
c.	6-Cl
d.	5,6-Benzo
e.	7,8-Benzo

321

Thiol Group
Moderate antipyretic and analgesic activities were shown by pyrimidine derivative [322a−e] with thiol groups at position-2 of rings as related to −OH-derived compounds.[78]

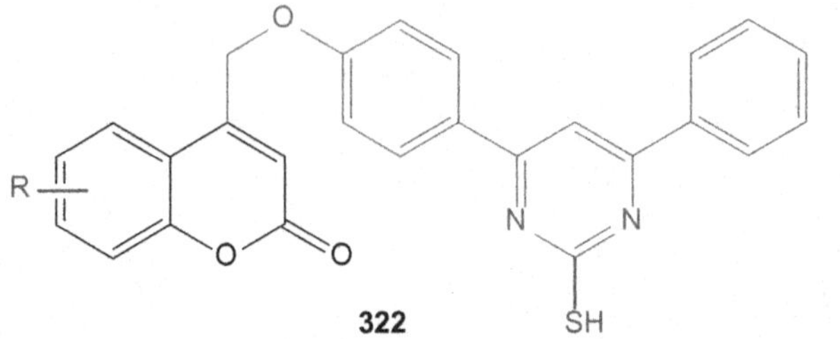

322	R
a.	6-CH3
b.	7-CH3
c.	6-Cl
d.	5,6-Benzo
e.	7,8-Benzo

322

6.1.3 Functional Group at C$_5$ Position
6.1.3.1 Small Group
A methoxy group [323] completely removed the TNF-α inhibitory effect. Some small group such as fluorine [324] is well accepted at this position.[75]

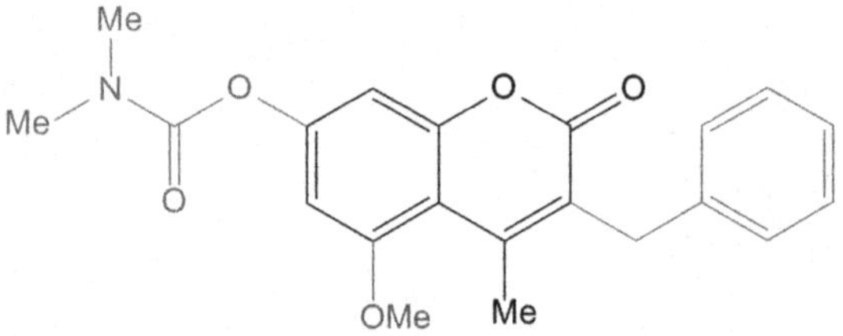

323

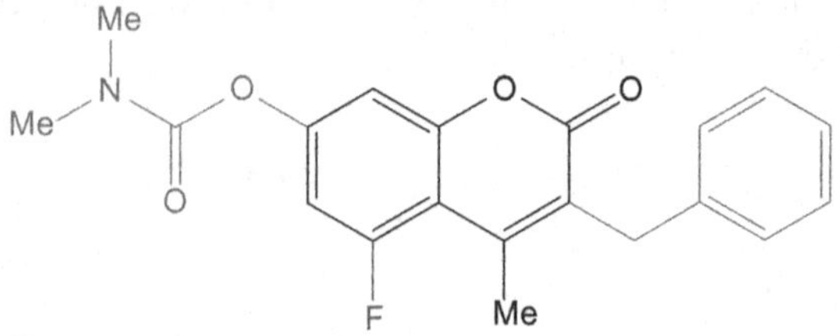

324

Compound [325] with a side chain at the C-4 position as well as another two methoxy groups in C-5 and C-8 positions exhibited IC50s value of 450 nM.[77]

325

6.1.3.2 Large Group

When the side chain of phenoxybutoxy was at the C-5 position of coumarin [326], it exhibited IC50 values of 150 nM.[77]

326

6.1.4 Functional Group at C_6 Position

6.1.4.1 Electron-Donating Group

The presence of an electron-withdrawing group such as CN, CHO, COOH, or NO_2 can enhance or weaken the TNF-α inhibitory effect. An electron-donating group such as MeO, as in [327], improved the inhibitory effect by threefold, analogous to the CN group.[75]

327

Alkyl Group

All alkyl substituents except methyl groups at the C-6 position caused a decrease in inhibitory activities.[75]

6.1.4.2 Halogen

The maximum improvement in TNF-α inhibitory activity was detected mainly for C-6 halo substituted compounds [328–330]. These compounds were 20−30 times more effective than the analogous non-substituted parent compound [328].[75]

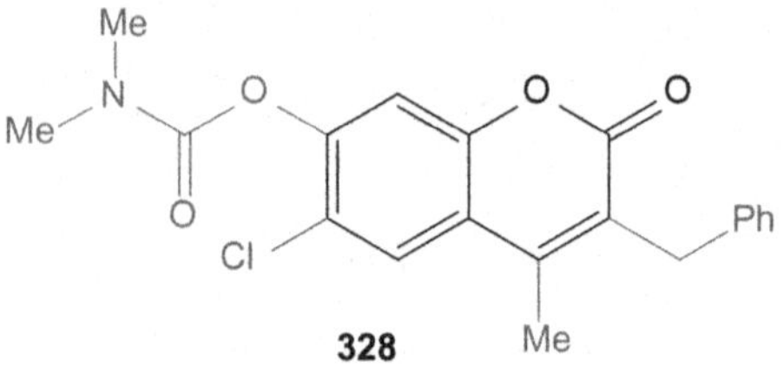

328

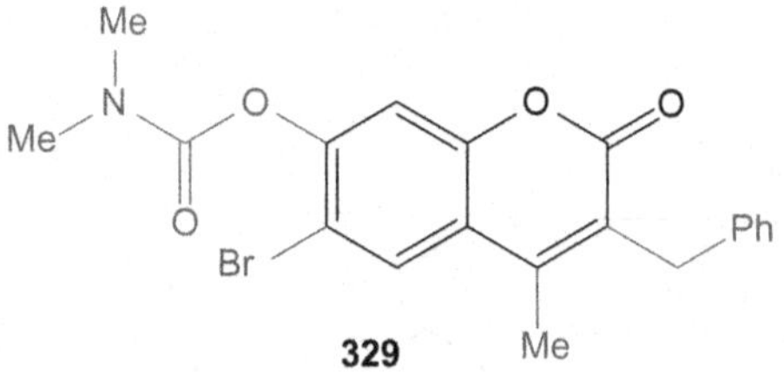

329

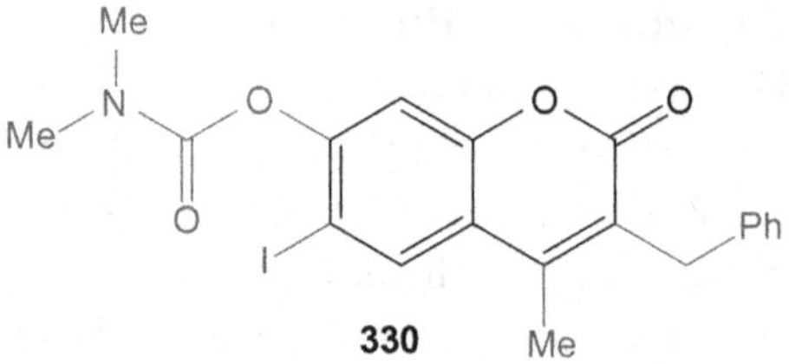

330

6.1.5 Functional Group at C_7 Position
6.1.5.1 Thio-Group Linkage
Thio-linkage between two aromatic moieties, as in compound [331], can have intense effects on 5-LO inhibitory activity.[76]

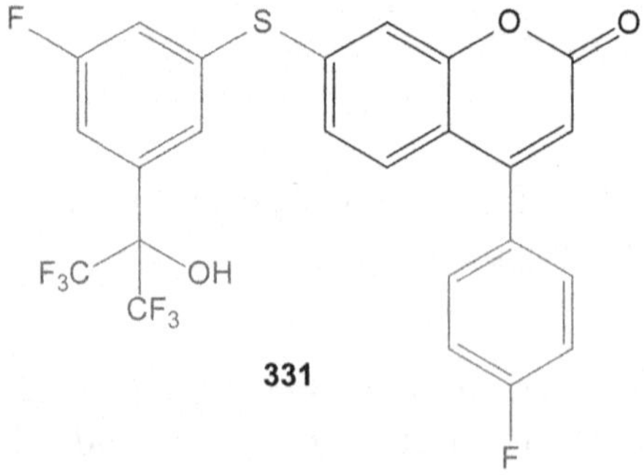

331

6.1.6 Functional Group at C_6 and C_7 Position
6.1.6.1 Ether Group

The presence of two small ether groups at the 6 and/or 7 position, with the methyl group at position 4, increases biological activity to develop a new anti-asthmatic drugs [332a−v].[79]

332	R_1	R_2	R_3
a.	H	OH	OH
b.	H	OMe	OMe
c.	H	OEt	OEt
d.	H	OPr	OPr
e.	H	OAc	OAc
f.	H	OCOPh	OCOPh
g.	H	OCH_2CO_2Et	OCH_2CO_2Et
h.	H	H	OH
i.	H	H	OMe
k.	H	H	OEt
l.	H	H	OPr
m.	H	H	OAc
n.	H	H	OCOPh
o.	H	H	OCH_2CO_2Et
p.	Me	H	OH
q.	Me	H	OMe
r.	Me	H	OEt
s.	Me	H	OPr
t.	Me	H	OAc
u.	Me	H	OCOPh
v.	Me	H	OCH_2CO_2Et

332

6.1.7 Functional Group at C_8 Position
6.1.7.1 Alkoxy Substituents

The psoralen derivative substituted with the side chain at C-8 position as compound [333] showed an IC50 value of 1 mM.[77]

333

Compound [334], having a side chain in the C-8 position, and further groups, such as methoxy group at the C-5 position and methyl substituents at the C-4 position, exhibited IC50s of 318 nM.[77]

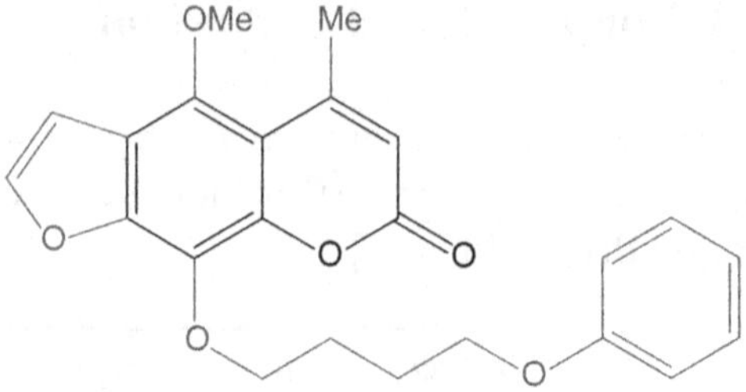

334

6.1.8 Furanocoumarin
6.1.8.1 Linear Furanocoumarins
Linear furanocoumarins, derivatives with oxyprenyl residues, showed moderate I_{GABA}.[80]

Hydroxyl and Methoxy Group
Furanocoumarins with hydroxyl (bergaptol) [335] or methyoxy groups [336, 235] such as bergapten and isopimpinellin do not exhibit any I_{GABA}.[80]

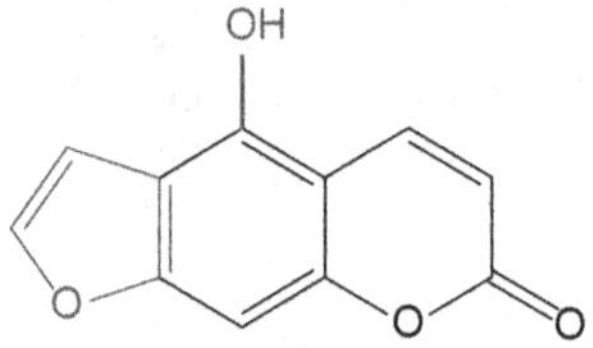

335

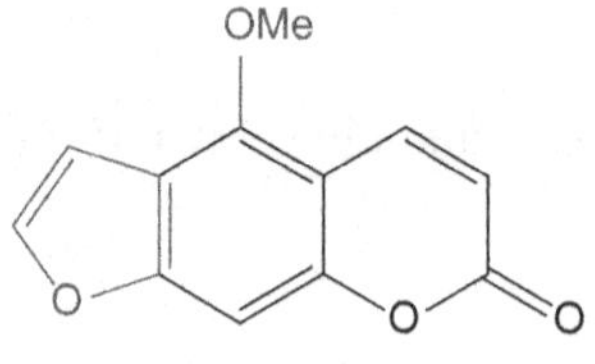

336

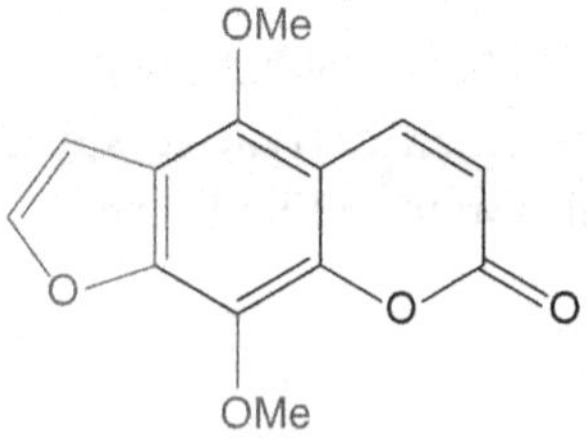

OMe
235

The gamma-aminobutyric acid type A receptor (GABA$_A$) is an inter-mediated ligand that functioned for fast neuronal signal transmission through gated ion channels. Coumarin derivatives and osthole are known to exert anticonvulsant activity.

Epoxy Group

The epoxy group enclosing oxypeucedanin [337] prompted the strongest promising effect. But at the same concentration, heraclenin [338] persuaded a reduced amount of I_{GABA} stimulation.[80]

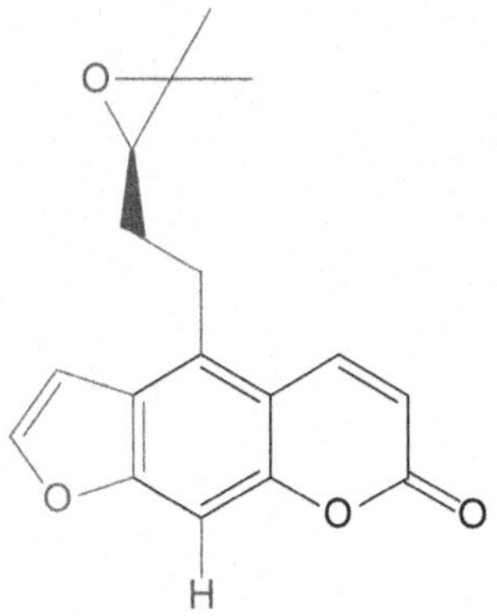

337

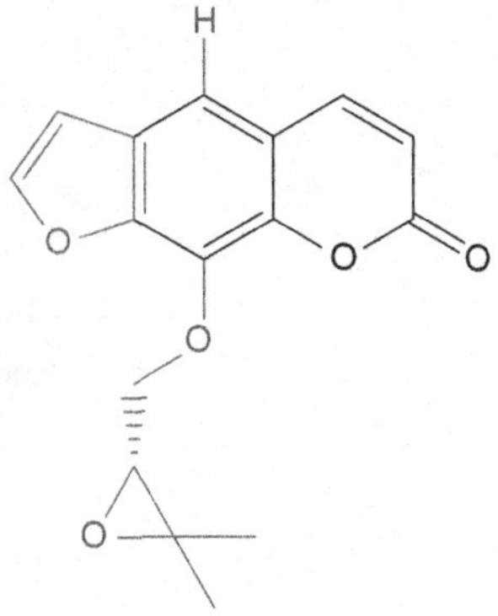

338

Oxyisopentenyl Group

Furanocoumarins using an oxyisopentenyl residue such as imperatorin [235], isoimperatorin [191], and phellopterin [236] revealed better I_{GABA} activity by $54 \pm 13\%$, $34 \pm 6\%$, and $57 \pm 4\%$, respectively.[80]

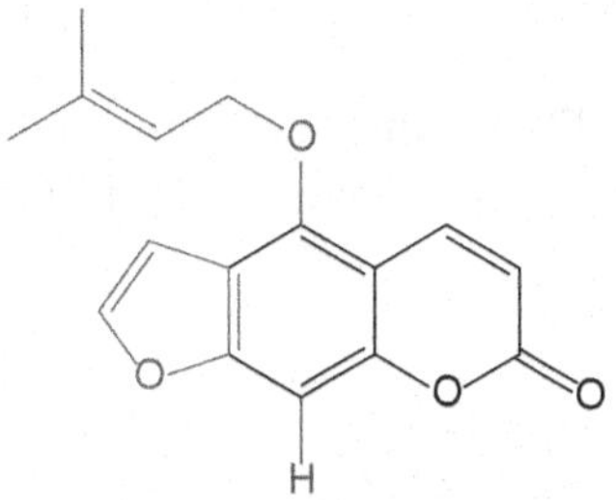

191

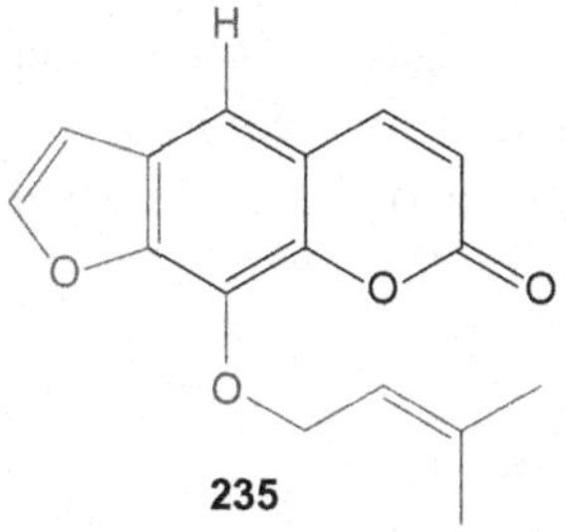

235

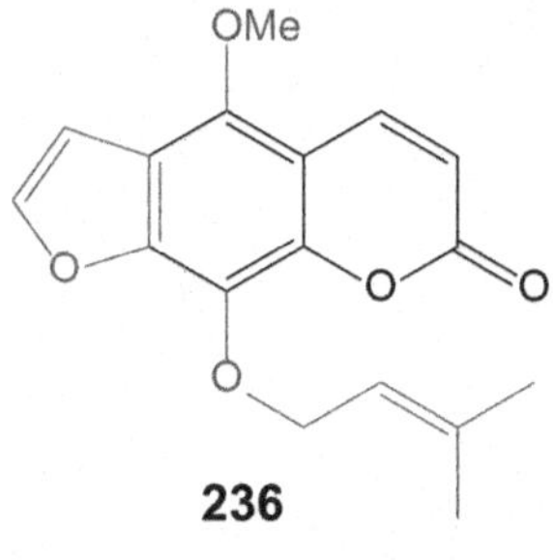

236

Oxygeranylated or Ester Derivatives

Furanocoumarins with bulkier residues such as oxygeranylated bergamottin [238] and the ester compound, ostruthol [339], did not exhibit
any improved I_{GABA} activity.[80]

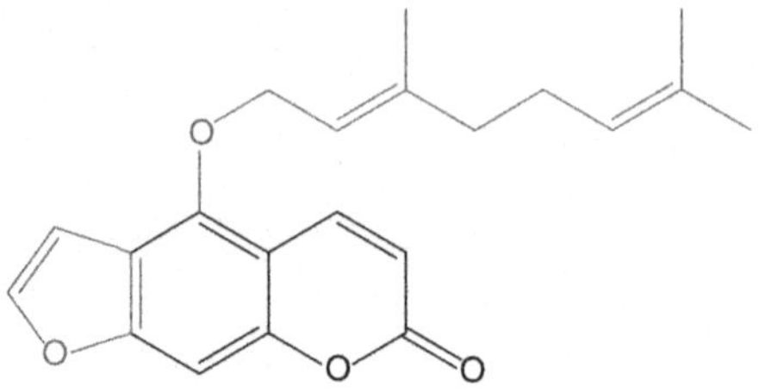

238

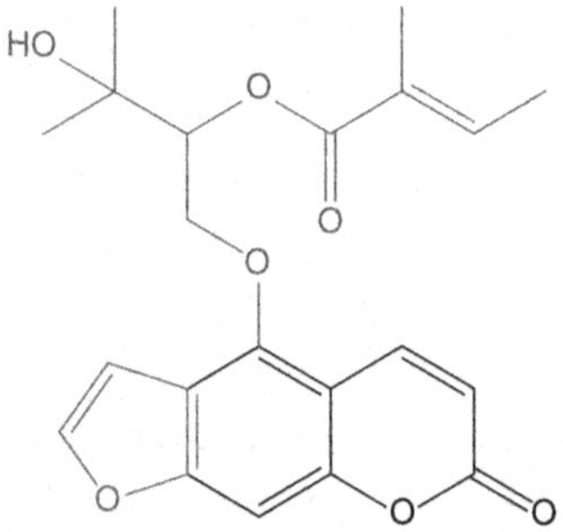

339

6.1.8.2 Angular Furanocoumarin

Angular furanocoumarin, isobergapten [340], displayed no activity,
whereas two times methoxylated derivative, pimpinellin [341], improved
I_{GABA} by $65 \pm 5\%$.[80]

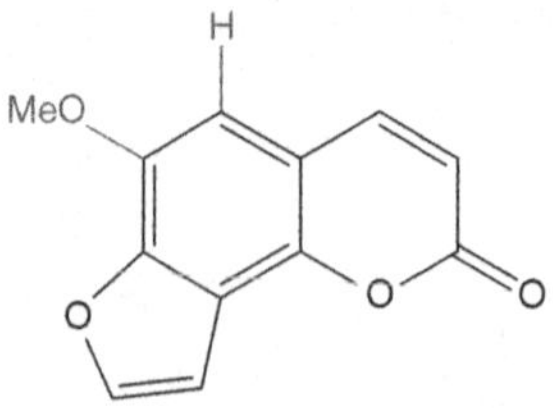

isobergapten
340

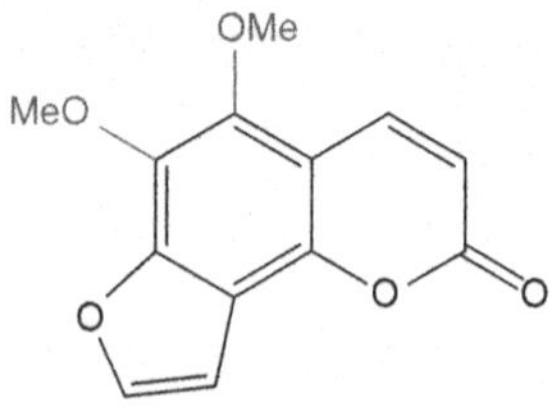

pimpinellin
341

CHAPTER 7

Antioxidant Agents

V. Rajeswer Rao
Department of Chemistry, National Institute of Technology, Warangal, India

7.1 INTRODUCTION

An *antioxidant* is the substrate that prevents the oxidation of molecules inside a cell. It is a well-known chemical process that allows the removal of electrons or hydrogen from a substance. Free radicals are produced during the biological oxidation reaction. Because the radicals are reactive, they start the chain reaction simultaneously. This can lead to the damage or even the death of a cell. Hence, antioxidant agents are capable of terminating a chain reaction by eliminating free radical intermediates, as shown in Figure 7.1. They perform the antioxidant behavior by being oxidized, hence antioxidants can be considered reducing agents. Some examples are ascorbic acid, thiols, or polyphenols. Antioxidants are commonly used as supplements in food (such as in Figure 7.2) and have been examined for inhibition of diseases such as heart disease and cancer.

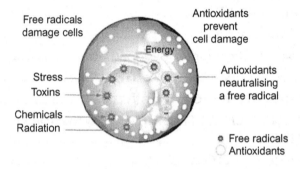

Figure 7.1 *Oxidation reaction in cell.* www.stapleton-spence.com/nutrition/antioxidants-and-free-radicals/.

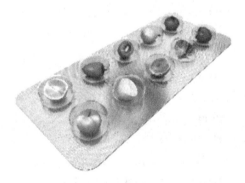

Figure 7.2 *Antioxidants.* www.4bestdietpills.com/wellness-advantages-of-dietary-pills-with-antioxidant/.

7.2 COUMARIN NUCLEUS

The antioxidant activity of pyrone ring of coumarin nucleus is briefly given in Figure 7.3.[81]

7.2.1 Dihydrocoumarin

Simple coumarin [1] and dihydrocoumarin [342] have no hydroxyl group substituted on the benzene ring of coumarin moiety were exhibited very small effects on the enzyme xanthine oxidase.[82]

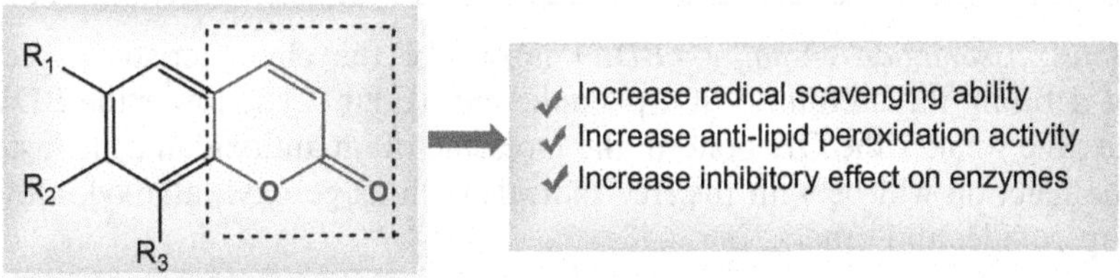

Figure 7.3 Importance of pyran-2-one nucleus on antioxidant activity.

7.2.2 Number and Position of Hydroxyl Group

Free-radical scavenging activities are related to number as well as position of the hydroxyl group on the benzene ring of the coumarin moiety. Furthermore, C-7, C-4, and C-2 substituted hydroxylated coumarin increase the activity.[40] It was also revealed that the interaction among the water particle and the C-4 hydroxyl group is essential for binding coumarin with the molybdopterin domain of xanthine oxidase.[82]

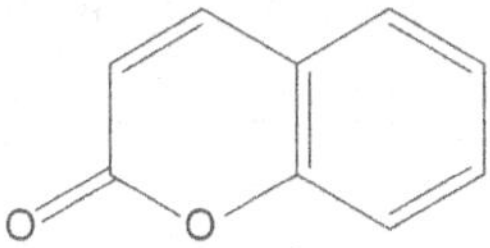

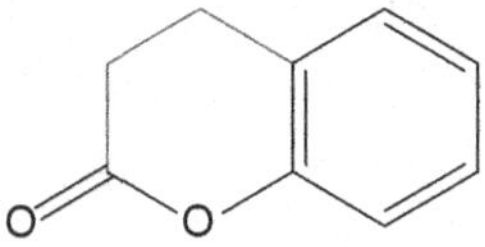

7.3 SAR BY BOND DISSOCIATION ENTHALPY

A phenol in conjugation with coumarin, O−H bond dissociation enthalpy for phenol was deliberated as 82.83 kcal/mol, so conjugation increased the oxygen-hydrogen bond dissociation enthalpy about 1 kcal/mol. The electron-donating group plays an important role in stabilizing the radical center and in reducing the O−H bond dissociation enthalpy, and vice versa.[83]

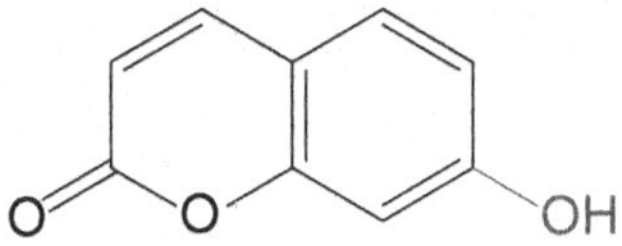

100

> *Bond dissociation enthalpy (BDE)* is one of the characteristic aspects of defining the usefulness of a phenolic antioxidant. The lower the BDE, i.e., the weaker the OH bond in the molecule of an antioxidant, the faster the reaction will be with the free radicals such as peroxyl, alkoxyl, alkyl, superoxide, and others.

The change in BDE of coumarins [100] and [343] (21.16 kcal/mol) entitles that ortho-OCH_3 decreases the BDE effect.[83]

343

Coumarin with resorcinol possesses the electron-withdrawing effect of meta-OH, the BDEs for two O—H of coumarin [344] cannot be reduced effectively hence, cause for low radical-scavenging activity.[83]

344

The catecholic radical formed due to catechol can be stabilized by the electron-donating effect, shown by the OH group present at the ortho position of the benzene ring as well as intramolecular hydrogen bonding. The bond dissociation enthalpy for conformation "down" [345] is 1 kcal/mol less than the conformation "up" [346].[83]

down

345

up

346

The O–H BDEs for up conformations [**348, 350**] is 4 kcal/mol higher than down conformations [**347, 349**] due to intra-molecular interactions. In up conformations hydroxyl group is protected against H-atom-abstraction ability. Thus, herein higher O–H BDEs are more efficient to scavenge radicals.[83]

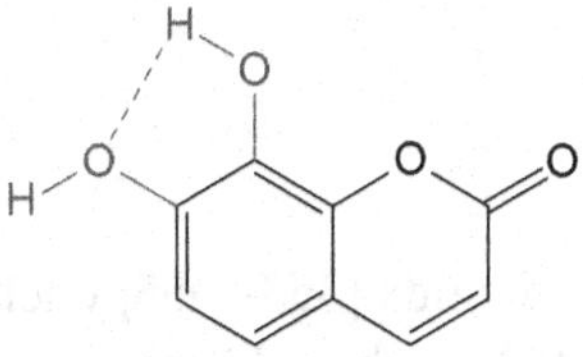

down
347

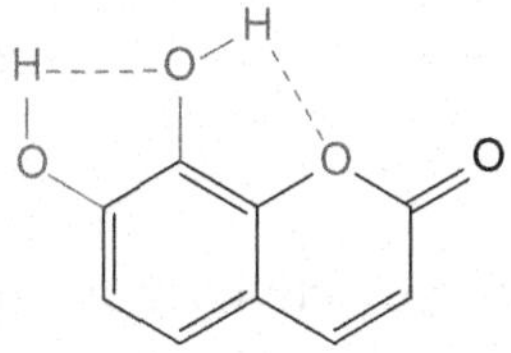

up
348

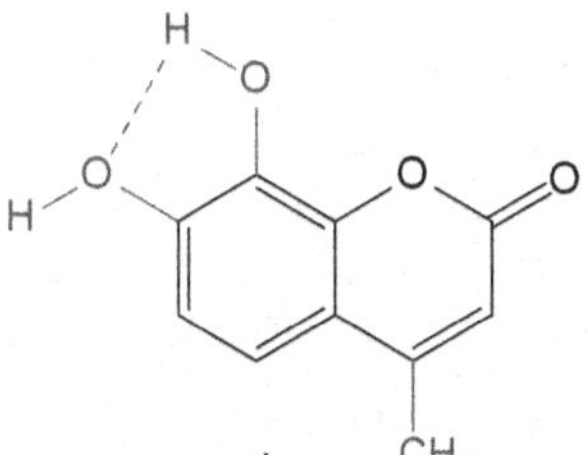

down CH₃
349

up CH₃
350

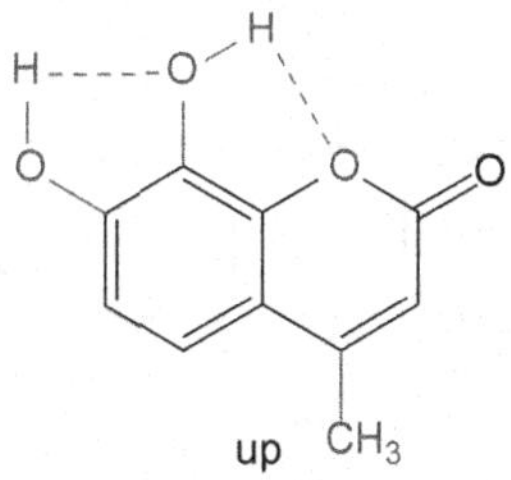

7.4 RADICAL SCAVENGING ACTIVITY

The maximum *in vitro* activity with the reference values of butylated hydroxyl toluene (BHT) and ascorbic acid (Asc), structurally promising residues, were acetyl groups of propene derivative [**351**].[84]

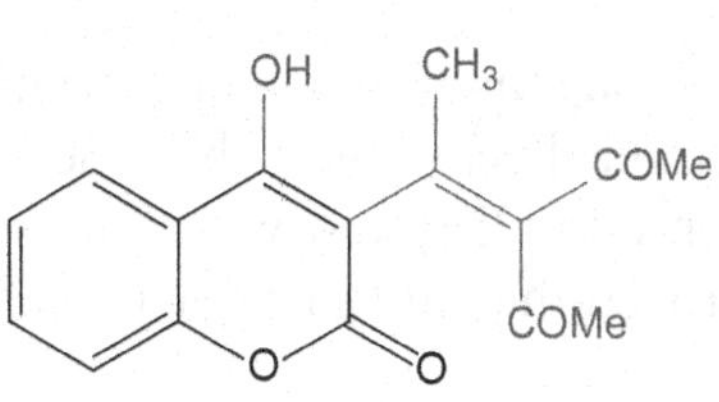

351

Scavenging potential in a small manner was decreased with the substitution of double carboxyethyl substitution such as in the compound [352] of the prop-1-en moiety.[84]

352

Reduced activity was observed for compounds [353–355] enclosing carboxymethyl groups or trans substituted carboxyl groups, using numerous cis-conformation. The cis-acetyl pharmacaphore established scavenging potential more in [353] regarding [354].[84]

353   **354**

355

The presence of strong hydrogen donor capability illustrates the most potent DPPH radical scavenger; thus the *N*-thiazole motif connected with *p*-SO$_3$H group [356] and with the existence of the further OH group [357] improved the antiradical activity.[84]

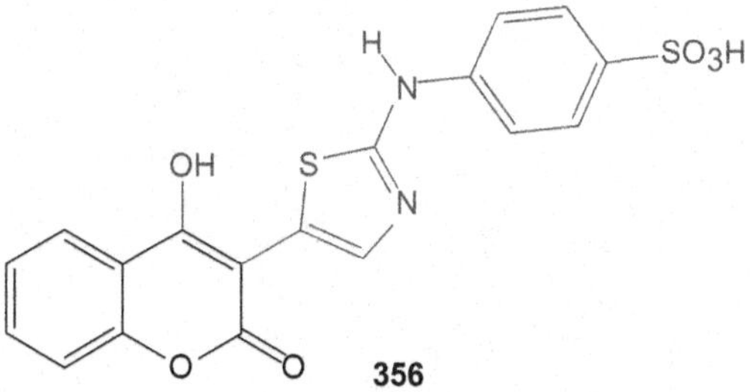

356

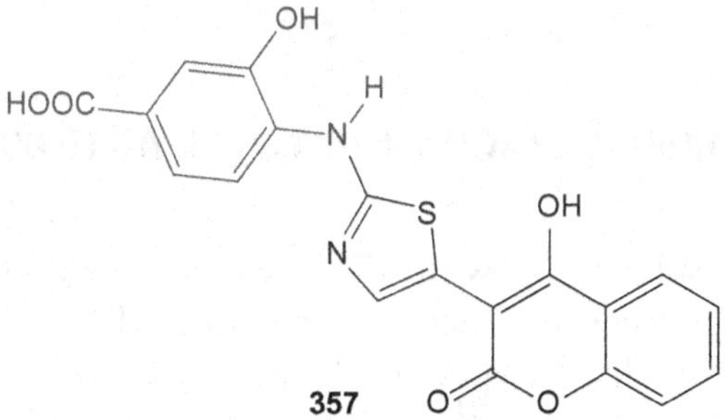

357

The m-NO$_2$, *N*-thiazole derivative [**358**] had a complete scavenging effect after 60 minutes of testing.[84]

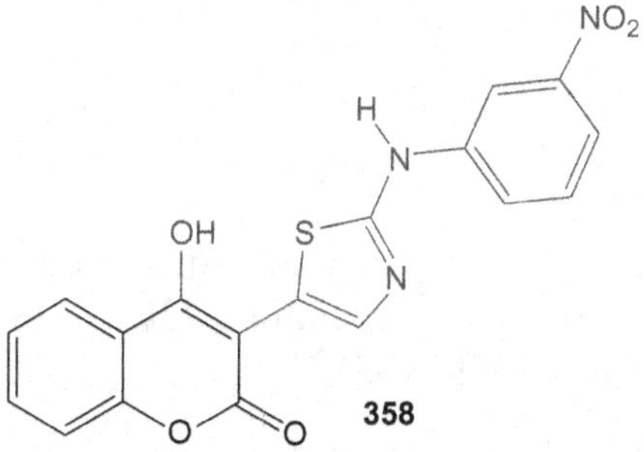

358

The thiazole derivative with *N,N*−diethyl [**359**], *N*-tolyl [**360−361**], and *N*-naphtyl groups exhibited the absence of activity.[84]

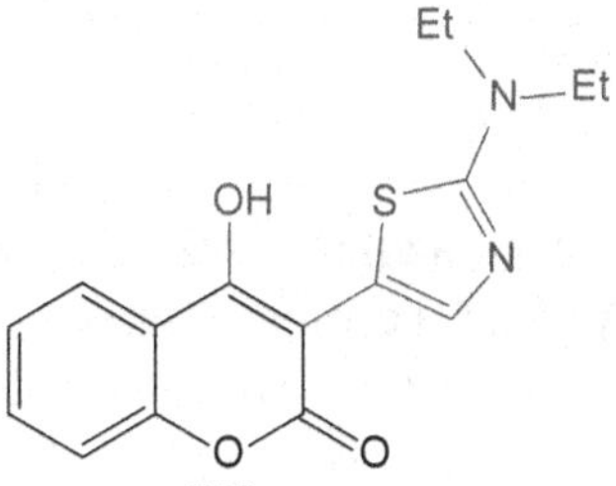

359

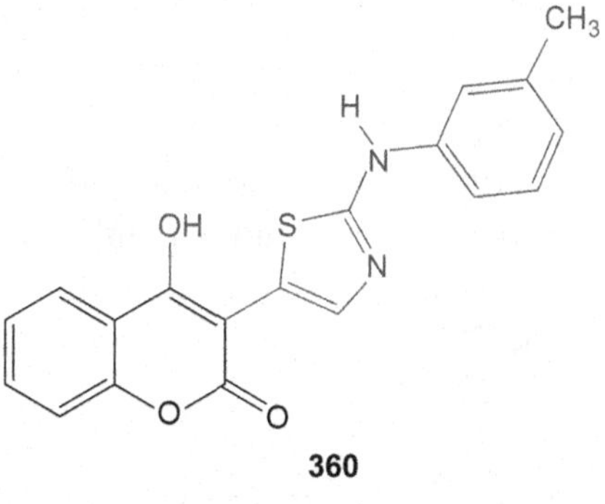

360

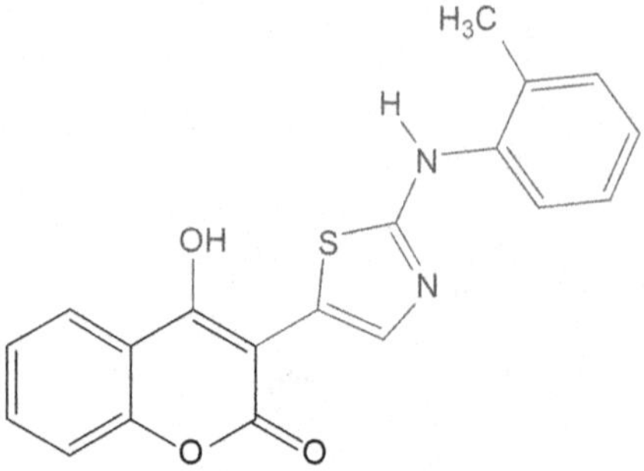

361

7.5 ANTIOXIDANT CAPACITY (OR TOTAL ANTIOXIDANT CAPACITY)

The coumarin substituted with (*E*)-2-cyano-3-methylbut-2-enoic acid [362] was the furthermost active compound. The *o*-hydroxybenzoic acid [357], *N,N*-diethyl [359], and tolyl [360–361] derivatives were also offered prominent potential.[84]

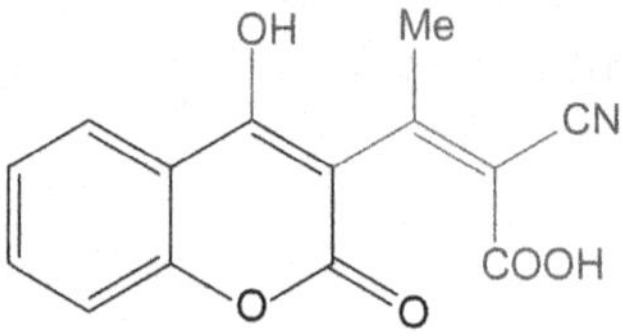

362

The decreasing order of the antioxidant activity of those compounds derived from compound [363] were as follows: 362 > 351 > 353 > 363 > 355 > 354.[84]

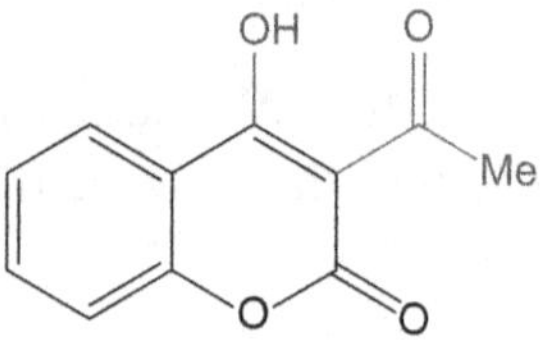

363

Within the *N*-thiazole derivatives, *N*-thiazole *p*-sulfonic acid-substituted compounds [356] were perceived as the most effective individuals through the maximum *total antioxidant capacity* (TAC) value.[84]

7.6 CHELATING ACTIVITY

The determination of chelating activity was achieved through the complex formation between Fe^{2+} and ferrozine. This red color complex is disrupted by chelating agents, with the consequence of reduction of color. 4-thiazolidinones [95, 364] showed the significant enhanced iron-chelating activity compared with the thiosemicarbazides [365], and particularly those complexes with alkyl groups on a 4-thiazolidinone ring. Compounds displayed iron-chelating activities, with 62.2%, 46.6%, and 4.54%, respectively, compared with EDTA disodium salt (97%).[37]

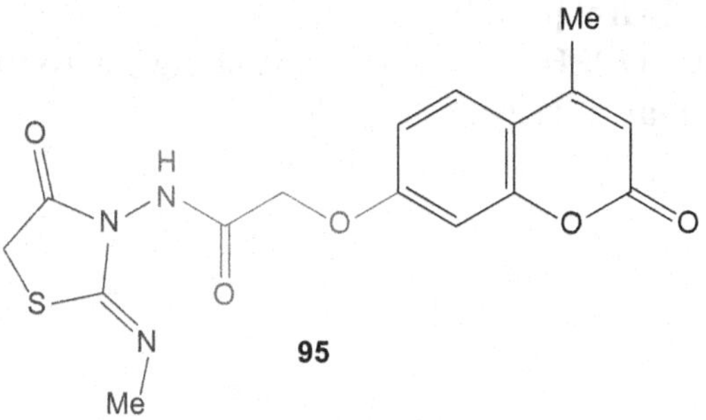

95

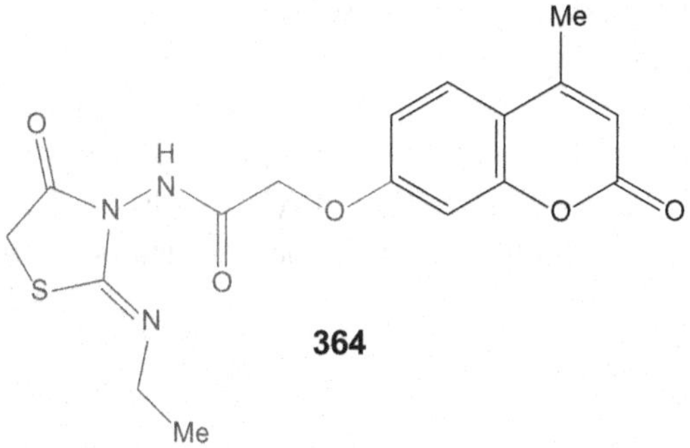

364

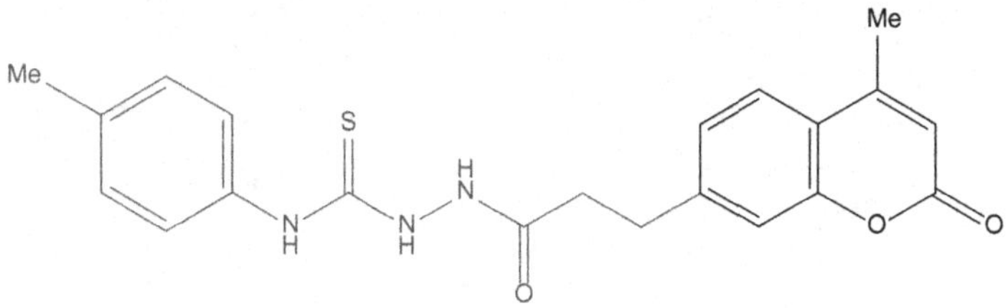

365

7.6.1 Functional Group at C₄ Position

7.6.1.1 Methyl Group

The CH_3 group at the C-4 position, as in simple coumarin [366], does not cause any change in oxidative activity.[85]

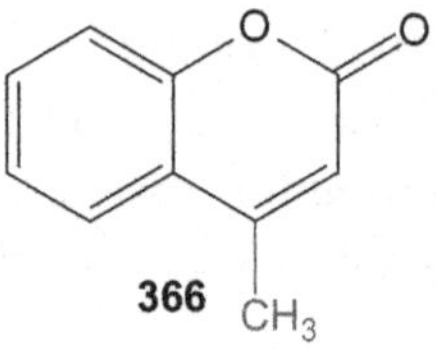

366

7.6.1.2 Hydroxyl Group

A significant DPPH radical scavenging activity shown by 4-hydroxy coumarin derivatives [367].[40]

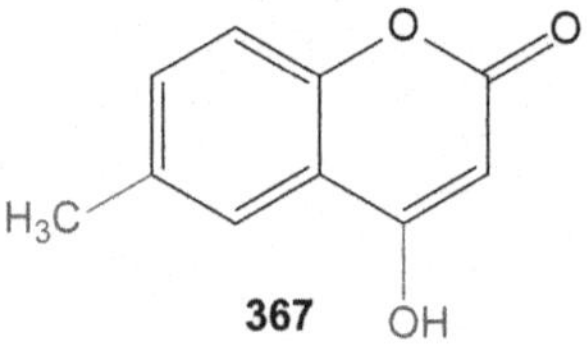

367

7.6.2 Functional Group at C₆ Position

7.6.2.1 Hydroxyl Group

Hydroxyl group substituents of the coumarin moiety would increase the antiradical activity; thus corresponding 3-acetyl-6-hydroxy-2H-1-benzopyran-2-one [368] as well as ethyl 6-hydroxy-2-oxo-2H-1-benzopyran-3-carboxylate [369] offered the best radical-scavenging action.[86]

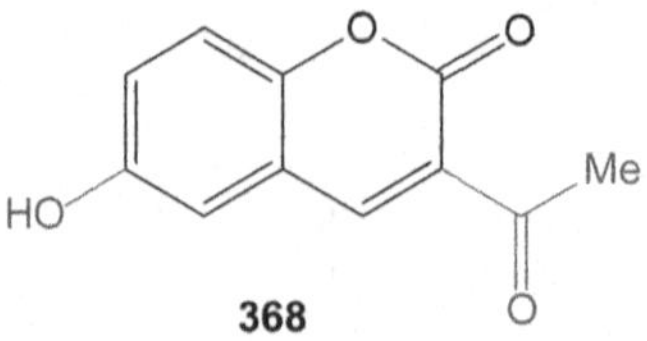

368

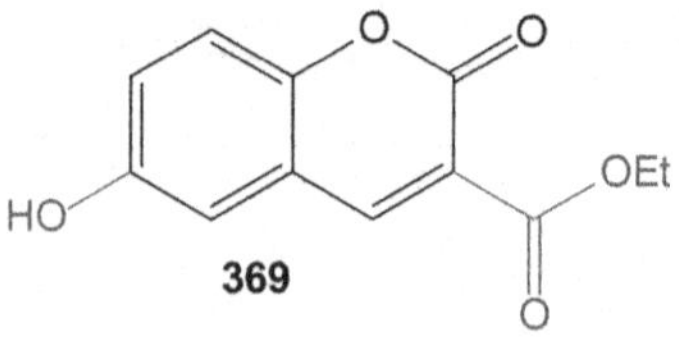

369

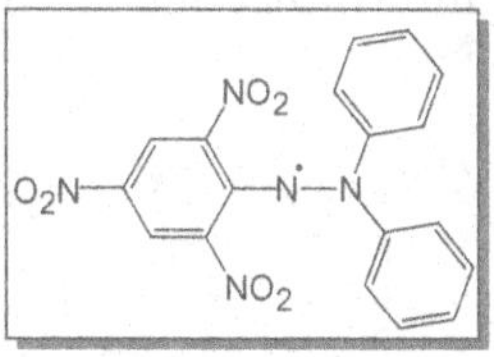

Figure 7.4 2,2-diphenyl-1-picrylhydrazyl.

DPPH is abbreviated for 2,2-diphenyl-1-picrylhydrazyl (Figure 7.4). It is a crystalline compound with stable free-radical fragments. DPPH is most commonly used in chemical reactions relating to radical formation to evaluate antioxidant activity and has one more application as standard of the locus and intensity of signals in electron paramagnetic resonance.

7.6.3 Functional Group at C_7 Position

7.6.3.1 Hydroxyl group

A significant DPPH radical-scavenging activity shown by 7-hydroxy coumarin [370–371] derivatives. Also substitution of hydroxyl derivative such as methoxy group in position-7, is also effective for the enhancement of antioxidant activity.[40]

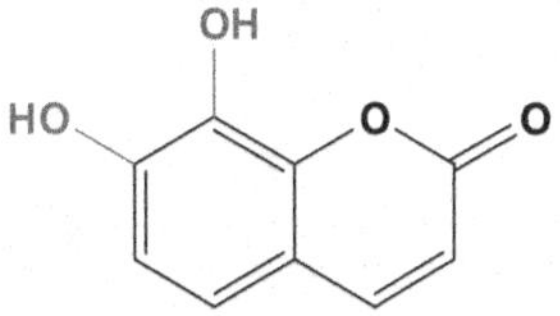

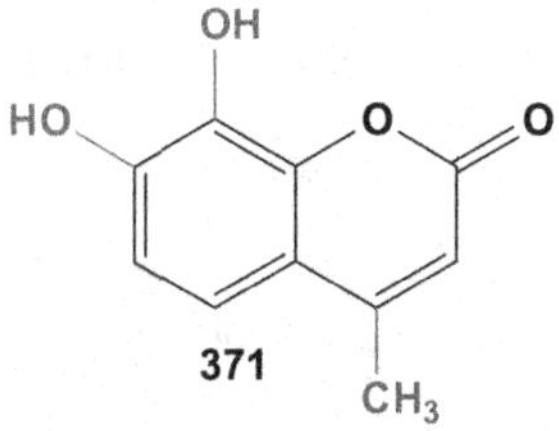

The coumarins possessing hydroxyl groups directly react with free radicals and inhibit the propagation of the chain reactions. Coumarin derivatives possessing hydroxyl groups are also testified as free radical scavengers and potent metal chelators. Thus they display effective antioxidant activity. Coumarin derivatives possess at least one hydroxyl group showing antioxidant activity.[40] The 7-hydroxyl substituents [146] form intermolecular hydrogen bonding with the active site of the enzyme.[82]

146

7.6.3.2 Thiosemicarbazide

With the analogous substituents, all thiosemicarbazides displayed better galvinoxyl- and DPPH-scavenging activity than 4-thiazolidinones, because thiosemicarbazides work as electron-donating groups to the free radical. The delocalization of electrons in the molecule forms a stabilized structure.[37]

Aryl and Alkyl Group

Compounds substituted with thiosemicarbazides and an aryl (phenyl) group displayed greater DPPH-scavenging effect compared to the alkyl substituents such as methyl and ethyl groups (EPR analytical method). However, when antioxidant activity evaluated through phosphomolybdenum, it displayed the alkyl group as more effective than aryl substituents. It was also revealed that p-methyl substitution on the phenyl ring [365] decreases and p-methoxy substitution [94] increases DPPH-scavenging effects of thiosemicarbazide derivatives.[37]

94

365

Sulphur Group

Thiosemicarbazide showed the effective antioxidant activity that is due to the presence of a C=S bond. The presence of phenyl groups also influences the antioxidant activity.[37]

7.6.4 Disubstituted Group

7.6.4.1 Functional Group at C_5 and C_7 Position

The presence of an OH group at C_5 and C_7 positions, such as in 5, 7-DHMC, exhibited a decrease in activity.[85]

344

5-Hydroxyeicosatetraenoic acid (5-HETE) is an endogenous acid signaling molecules synthesized from oxidation of 12-hydroxy-5,8,10-heptadecatrienoic acid and arachidonic acid in polymorphonuclear leukocytes. They apply multifarious control in numerous biological functions, primarily in inflammation or resistance and as messengers in the CNS. Coumarin-based derivatives are effective 5-hydroxy-6,8,11,14-eicosatetraenoic acid inhibitors.[85]

5-hydroxy-6,8,11,14-eicosatetraenoic acid

12-hydroxy-5,8,10-heptadecatrienoic acid

arachidonic acid

7.6.4.2 Functional Group at C_6 and C_7 Position

For effectiveness in the inhibition of the 5-HETE, the presence of two hydroxyl groups neighboring each other at positions 6 and 7 of the coumarin ring is essential. But oxidative ability decreases with the position of the OH group; as present in 6,7-DHMC [373] showed a decrease in the inhibitory effect.[85] Esculetin [372] and 4-methylesculetin [373] showed significant activity against molybdopterin moiety. This was expected because of interaction between the E802 residue and the 6-hydroxyl group.[82]

Esculetin
372

4-methylesculetin
373

7.6.4.3 Functional Group at C_7 and C_8 Position

5-HETE can be inhibited effectively by the two neighboring OH groups at C_7 and C_8 positions of the coumarin structure. It also decreases the oxidative ability shown by the 7,8-DHMC [371]. Monohydroxy coumarins, such as umbelliferone and scopoletin, also repressed the development of 5-HETE, but not very powerfully.[85]

371

CHAPTER 8

Anticoagulant Agents

Gudala Satish
Department of Chemistry, National Institute of Technology, Raipur, Chhattisgarh, India

8.1 INTRODUCTION

Coagulation is the usual process that takes place inside the human body whenever it is necessary. The process involves 14 different types of naturally occurring coagulating agents. Sometimes irregularity in the immune system causes unnecessary coagulation. In such cases, *anticoagulant agents* (Figure 8.1) are used to inhibit the clotting of blood or prevent remaining coagulates from increasing. These agents can keep harmful clots from forming in the heart, arteries, or veins. Clotting in blood can also be the reason for a heart attack. Anticoagulants are the medicines used *in vivo* thrombotic conditions. Food supplements that have blood-thinning properties are papaya, beer, garlic, pomegranate, fish oil, ginseng, ginger, green tea, onion, turmeric, and the like. Coumarin-based naturally occurring warfarin (Figure 8.2) is a well-known anticoagulating agent.

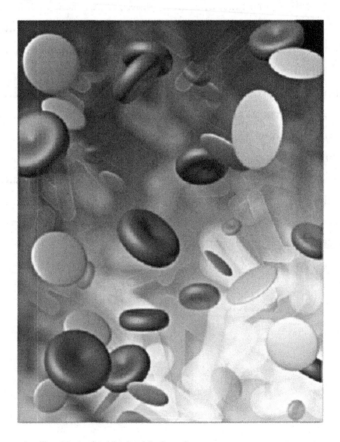

Figure 8.1 www.drgailhumble.com/tag/platelet-rich-plasma/.

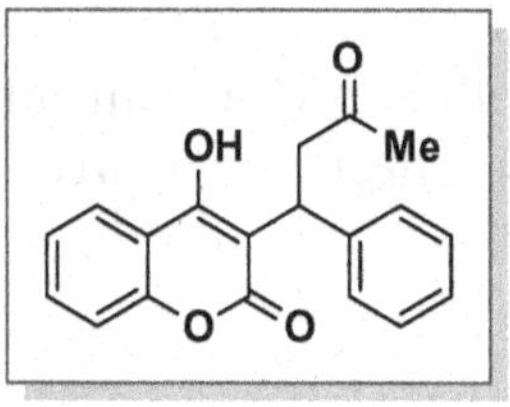

Figure 8.2 Warfarin (coumarin based) anticoagulant.

8.1.1 Functional Group at C$_3$ Position
8.1.1.1 Acetoxy Group
The maximum catalytic effect was detected in acetoxy coumarins without substitution at the C-3 position, such as 7-acetoxy-4-methylcoumarin (MAMC) and 7,8-diacetoxy-4-methylcoumarin (DAMC) [375].[87]

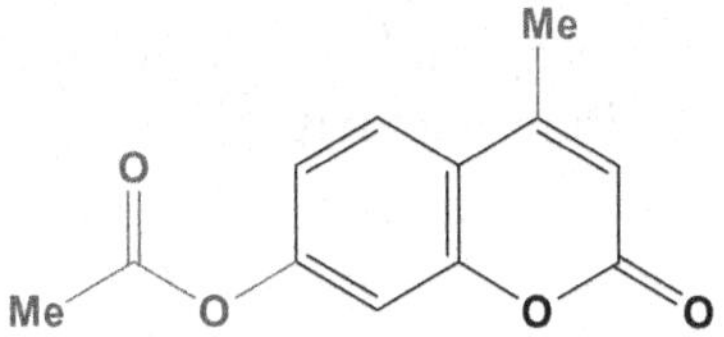

8.1.1.2 Alkyl Group
The presence of bulky alkyl substituents at position 3 of the coumarin nucleus decreases its capacity for acetyl group transformation to functional proteins. This demonstrates that a steric factor is responsible for decreased affinity of acetoxy groups for binding the active sites of the enzyme CRTAase.[87]

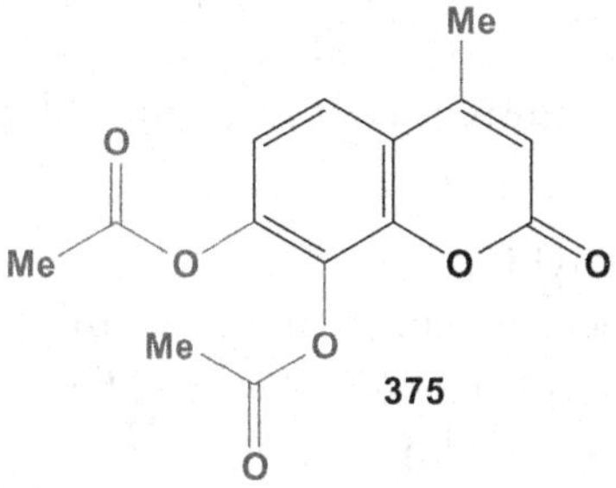

Calreticulin transacetylase (CRTAase) is a membrane-bound enzyme found in mammalian cells which catalyses the transfer of acetyl groups from polyphenolic acetates (PAs) to certain functional proteins via glutathione S-transferase (GST), nitric oxide synthase (NOS), and NADPH cytochrome-c reductase, causing the variation of their biological activities. The alkylated acetoxy coumarins are potent in inhibiting the catalytic activity of mainly GST and ADP-induced platelet accumulation through the activation of platelet nitric oxide synthase (NOS).

Heptyl and Nonyl Group

Thus, compounds [376] with propyl substituents at position 3 are responsible for better CRTAase effects, compared with the heptyl or nonyl substituted derivatives.[87]

376	R_1	R_2
a.	$n\text{-}C_3H_7$	H
b.	$n\text{-}C_7H_{15}$	H
c.	$n\text{-}C_9H_{19}$	H
d.	$n\text{-}C_3H_7$	OAc
e.	$n\text{-}C_7H_{15}$	OAc
f.	$n\text{-}C_9H_{19}$	OAc

376

8.1.1.3 Amine Group

Cyclic or Linear Group

Substitution of an amine group on cyclic or linear aliphatic side chains constantly leads to a diminishing in THR inhibitory activity.[88]

Phenyl Group

When a phenyl ring is substituted with an amine group, potent THR inactivators are furnished [378, 379]. Particularly meta substituted amine [378] is highly prominent.[88]

378

379

8.1.1.4 Guanidine Group

Side chains at the 3-position bearing 2-guanidino-ethyl and 4-guanidino-phenyl [377 a and c], respectively, have weak THR inhibitory activity.[88]

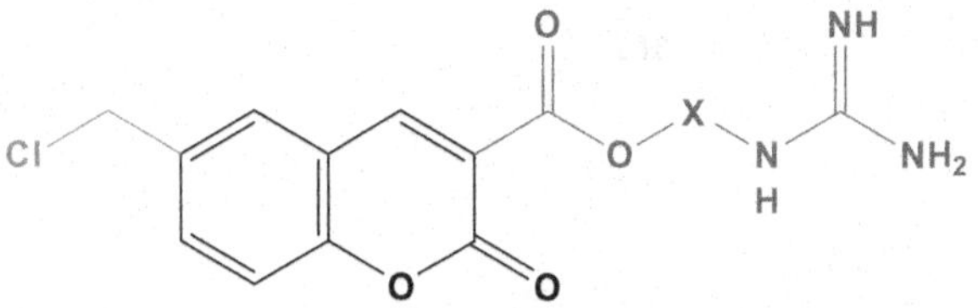

377	X
a.	2-ethyl
b.	3-propyl
c.	m-phenyl
d.	p-phenyl
e.	p-cyclohexyl
f.	4-piperidine

377

Thrombin (THR) is a trypsin-like serine protease, that plays an essential role in the process of haemostasis and thrombosis. Guanidine derivatives do not exhibit any THR inhibitory activity.

8.1.1.5 Pyridine Group

In the set of pyridine derivatives [380–384], compounds with the pyridine nitrogen in the 2'-position with mainly a chlorine atom in the 3'-position effectively inhibit THR [380].[88]

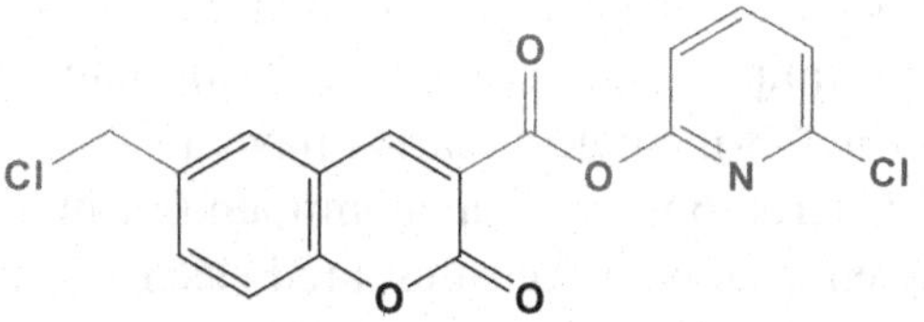

380

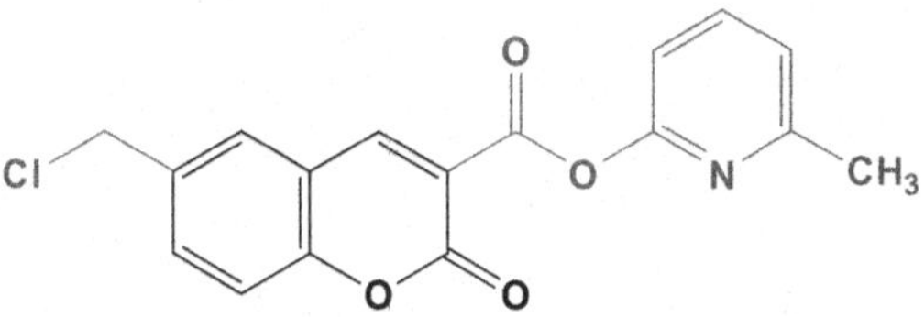

381

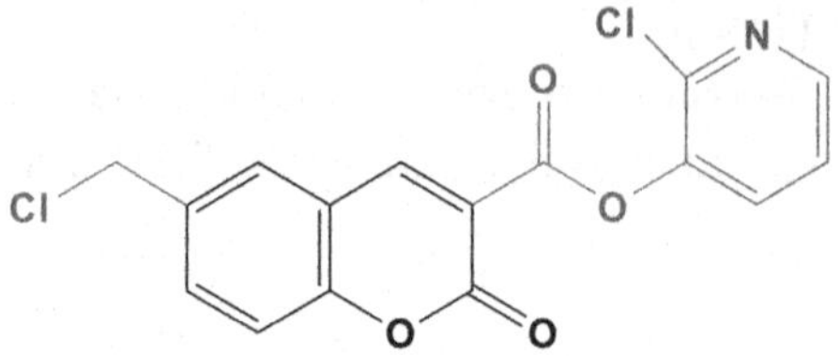

382

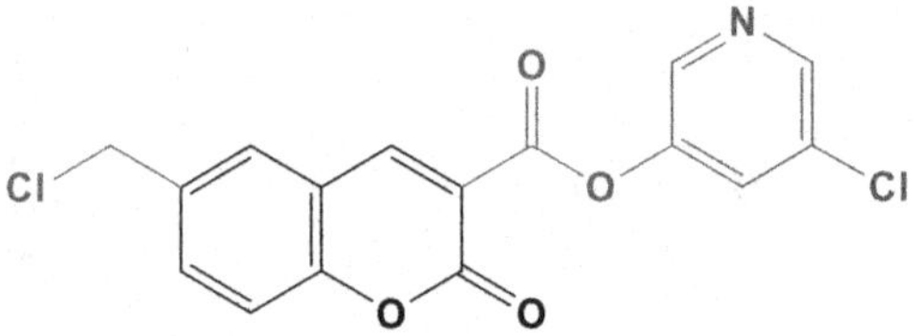

383

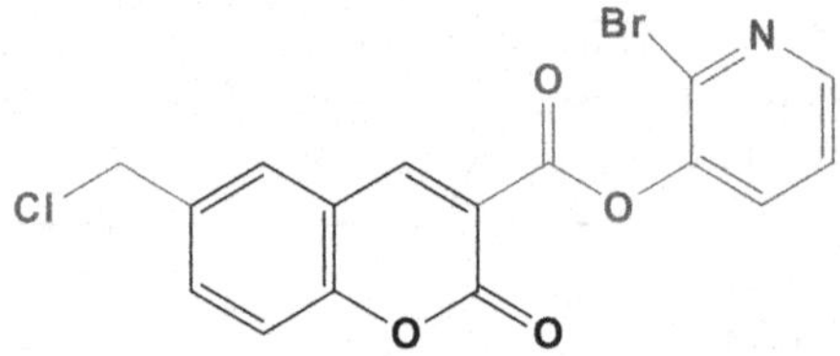

384

8.1.2 Functional Group at C$_4$ Position
8.1.2.1 β-enaminone and Piperazinyl Group

The β-enaminone moiety [385] and 1-piperazinyl group [386] are important for the anticoagulant effect (confirmed by the significant change among the IC50 values found for derivatives). Thus, it was expected that 4-(1-piperazinyl)-8-methylcoumarins have a 7-alkoxy group as well as appropriate heteroatoms so that compounds can interact with the active site of the enzyme with their unshared electron pairs.[89]

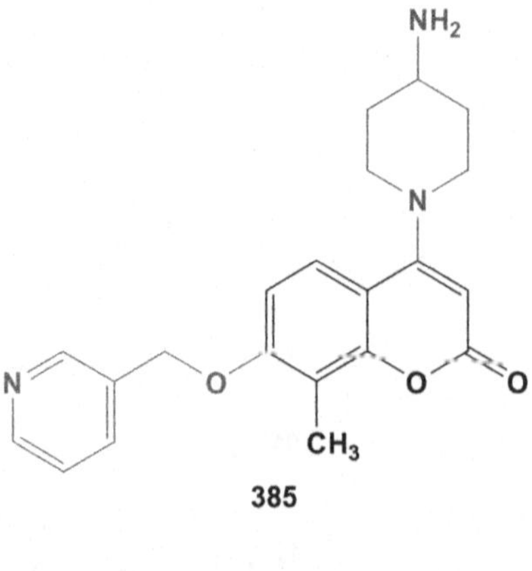

385

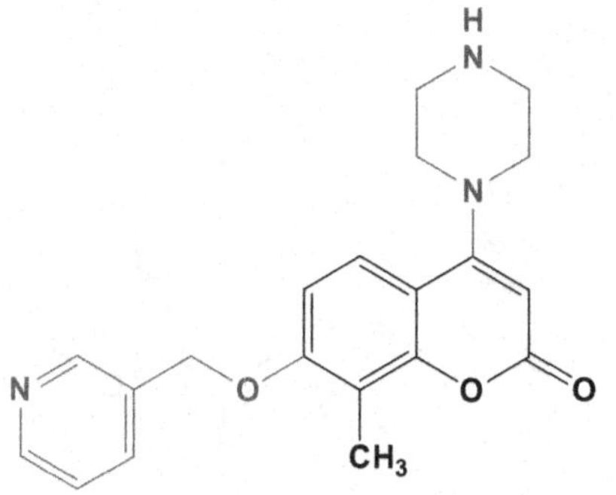

386

8.1.3 Functional Group at C_7 Position
8.1.3.1 Alkoxy Group
Effective anticoagulating activity was found in the 7-alkoxy substituted coumarin derivatives.[89]

Propoxy and Butoxy Group
Significant increases in the inhibitory activity of human platelet phosphodiesterase were obtained in the compounds by the isosteric substitution of a CH_2 moiety in the 7-butoxy and 7-propoxy group [387–388] with appropriate heteroatoms such as oxygen and sulphur.[89]

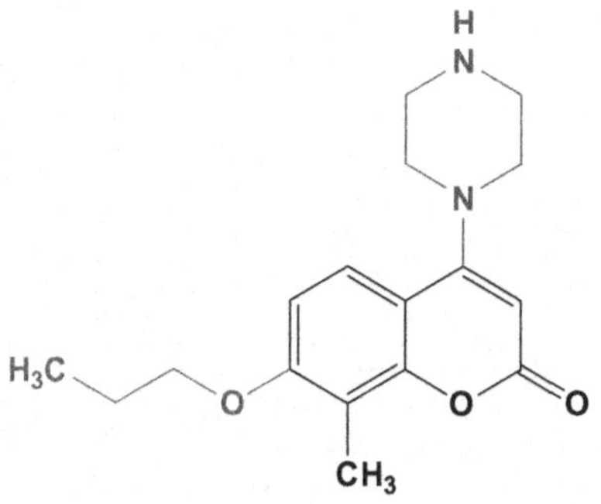

387

8.1.3.2 Amine Group
The substitution of the CH_3 group from the 7-propxy chain by a disubstituted NH_2 group [389–391] gave irregularly less effective derivatives, apart from the presence of the unshared electron pairs of nitrogen.[89]

388

389

390

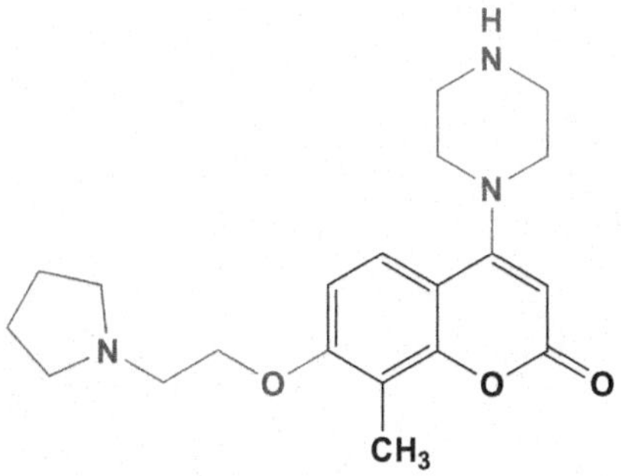

391

8.1.3.3 Phenyl Group

The substitution of a phenyl group [392] caused the delocalization of unshared electron pairs of oxygen atoms, which was the cause for lowering in activity of the compound.[89]

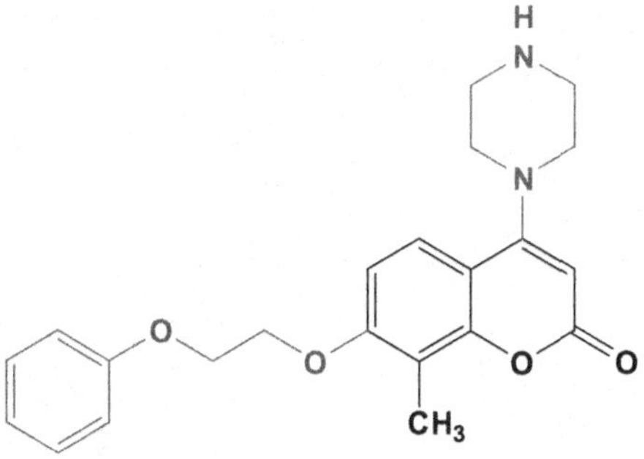

392

8.1.4 Functional Group at C₈ Position

8.1.4.1 Methyl Group

The methyl group at the C-8 position such as in [386] is important for the platelet anti-aggregating effect of compounds.[89]

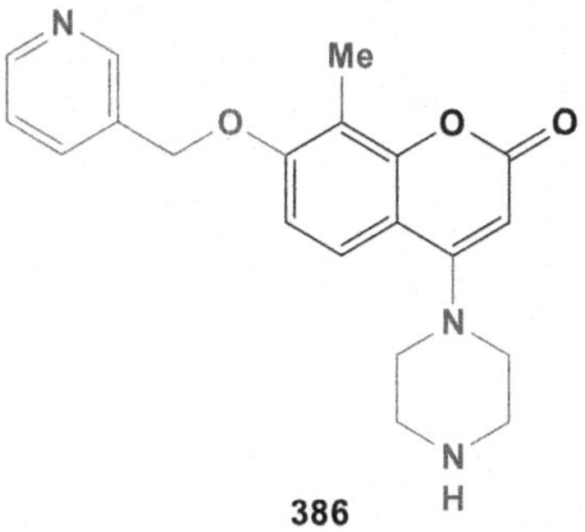

386

CHAPTER 9

Miscellaneous Agents

Archi Sharma

Department of Chemistry, National Institute of Technology, Raipur, Chhattisgarh, India

9.1 DOPAMINE INHIBATORY ACTIVITY

Dopamine-releasing activity is responsible for several disorders, such as induced cytotoxicity in PC12 cells, Parkinson's,[90] depression, and Alzheimer's diseases.[91]

9.1.1 Hydroxyl Group

Docking study and target binding site of protein kinase II B (CAMK2B) in association with the effect of released dopamine showed that a 4-hydroxy-2-*H*-chromen-2-one core [**393**−**396**] played an important role in the chief interaction with the active site over its 2-carbonyl group and 4-hydroxyl group.[91]

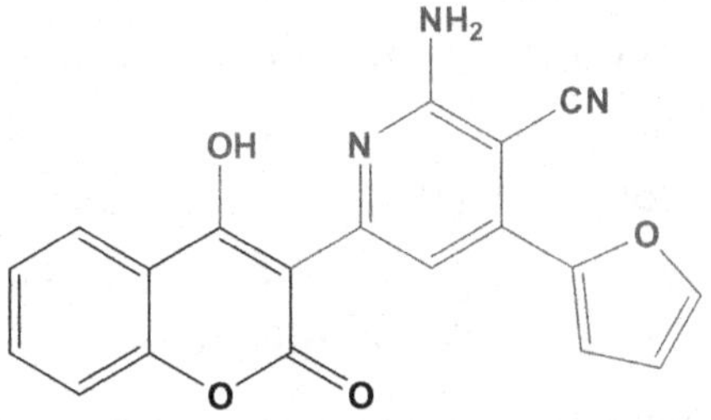

393

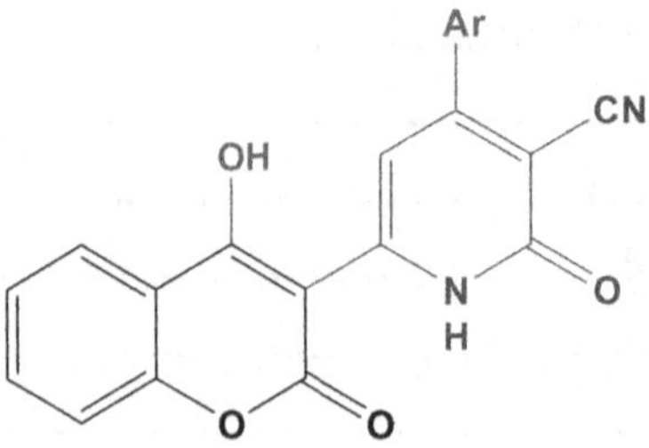

394

394	Ar
a.	C_6H_2-3,4,5-$(OCH_3)_3$
b.	3-Pyridyl
c.	2-furyl

396	Ar
a.	C_6H_2-3,4,5-$(OCH_3)_3$
b.	3-Pyridyl

9.1.2 Pyrazole Group

3-diaryl pyrazole derivatives are more effective compared with 3-monoaryl derivatives. Enhanced activity for *p*-methoxy 3-aryl or unsubstituted or *N*, 3-diaryl pyrazole was found related to their substituted trimethoxy derivatives. The compounds that had *N*-alkyl pyrazole groups were least active among all the derivatives.[91]

397	R	Ar
a.	H	$C_6H_4$4-(OCH_3)
b.	H	C_6H_2-3,4,5-(OCH_3)
c.	H	3-pyridyl
d.	$COCH_3$	3-pyridyl
e.	Ph	C_6H_2-3,4,5-(OCH_3)
f.	Ph	3-pyridyl
g.	$CSNH_2$	$C_6H_4$4-(OCH_3)
h.	$CSNH_2$	3-pyridyl
i.	CSNHPh	$C_6H_4$4-(OCH_3)
j.	CSNHPh	C_6H_2-3,4,5-(OCH_3)
k.	CSNHPh	3-pyridyl

9.2 DNA-PK INHIBITORY ACTIVITY

DNA-dependent protein kinase (DNA-PK) shows an important role in the damage responses of DNA by repairing the DNA double-stand break through the nonhomologous end-joining pathway.[92]

9.2.1 Steric Effect

At the C-6 position of coumarin, the substitution of the larger aryl [398–400] group showed decreased or complete loss of activity, which is expected due to some steric hinderance.[93]

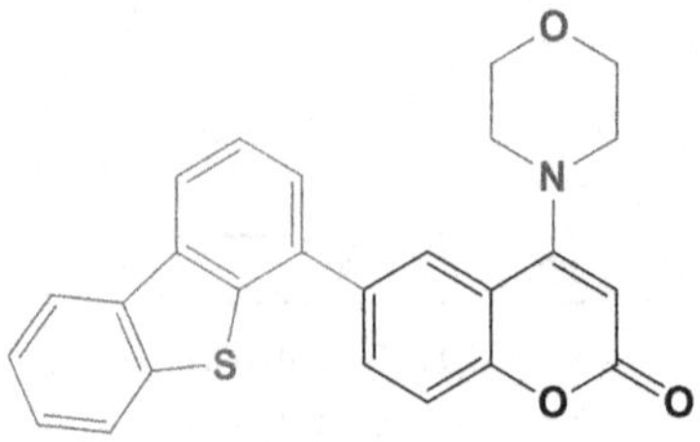

398

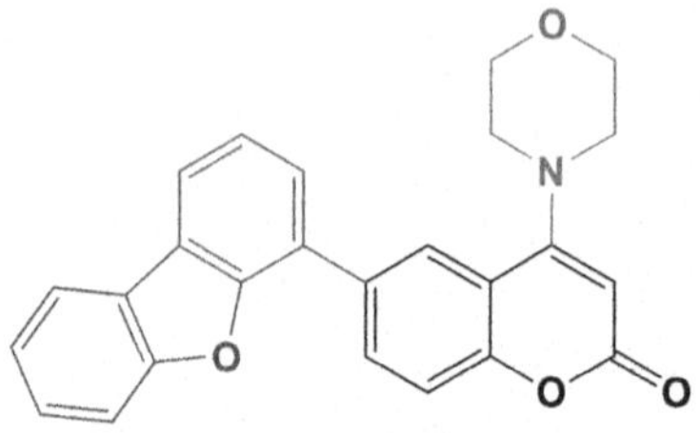

399

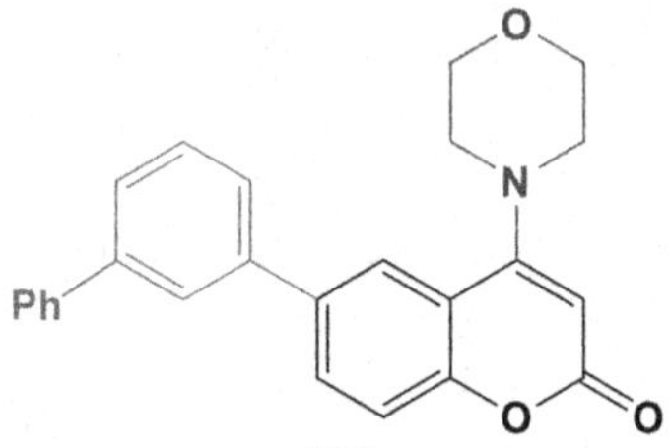

400

Similarly, decreases in potency were even more evident for the dibenzofuran-1-yl [**401**], Benzothiophen-2-yl [**402**], and dibenzothiophen-1-yl [**403**]. 7-substituted coumarins were 70 times less effective.[93]

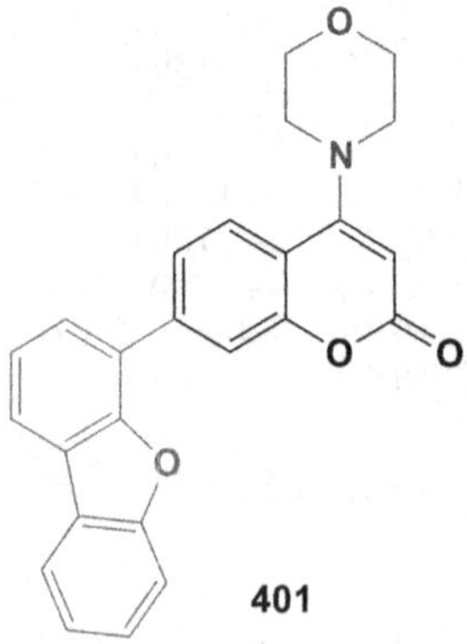

401

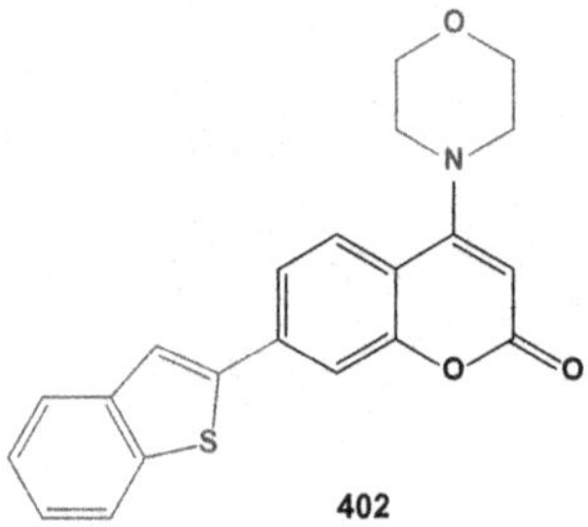

402

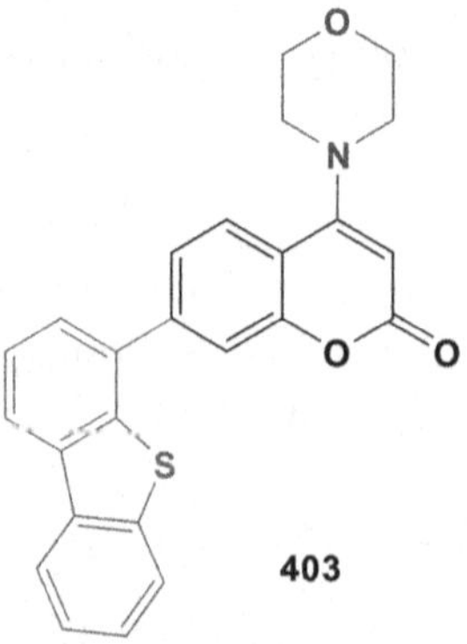

403

9.2.2 Functional Group at C-6 and C-7 Position
9.2.2.1 Thienyl Group

The 2-thienyl group at the 7-position of the coumarin [404] nucleus displayed a submicromolar effect, but the 6-substituted coumarin [405] was rather less effective.[93]

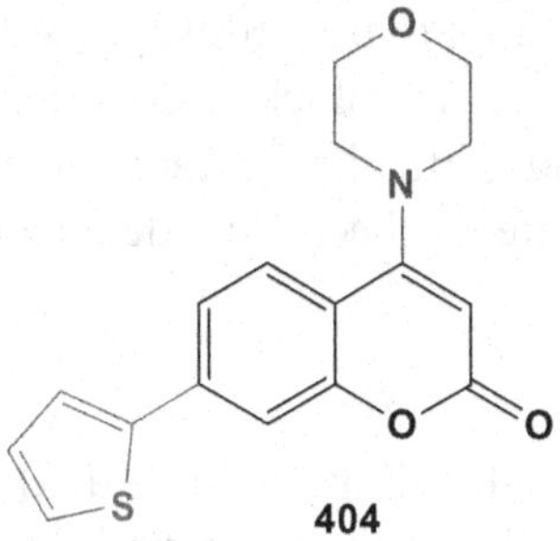

404

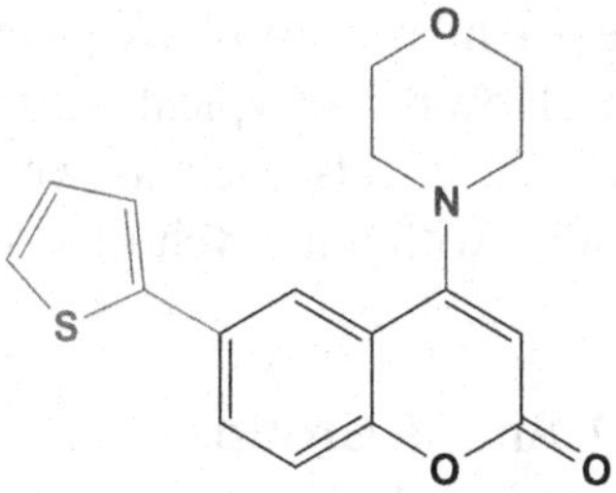

405

9.2.2.2 Methoxy Group or Phenyl Group

Coumarin rings substituted with methoxy [406–407] or phenyl groups [408–409] at the 6- or 7-position gave derivatives with equivalent DNA-PK inhibitory activity.[93]

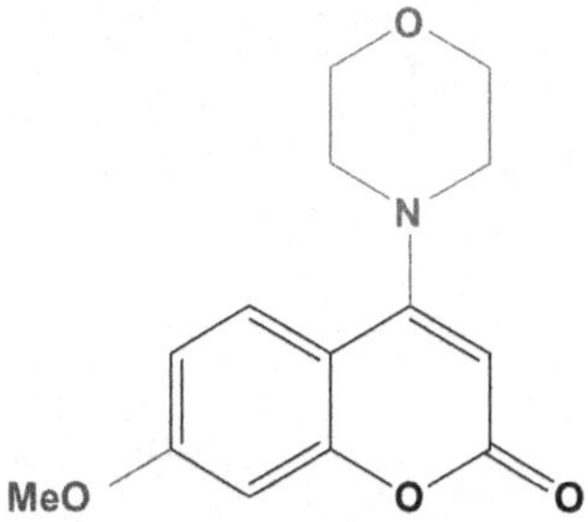

406

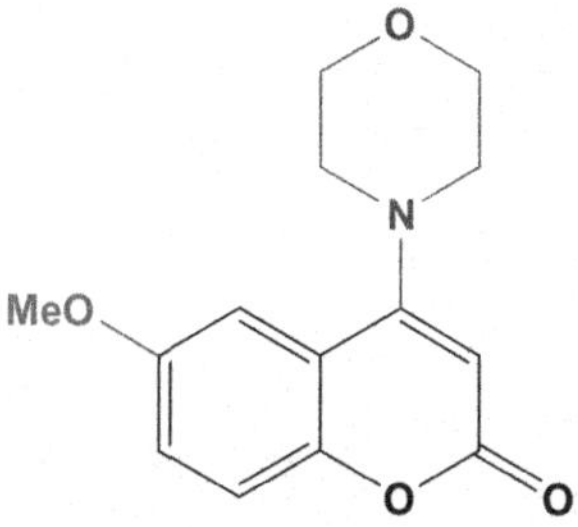

407

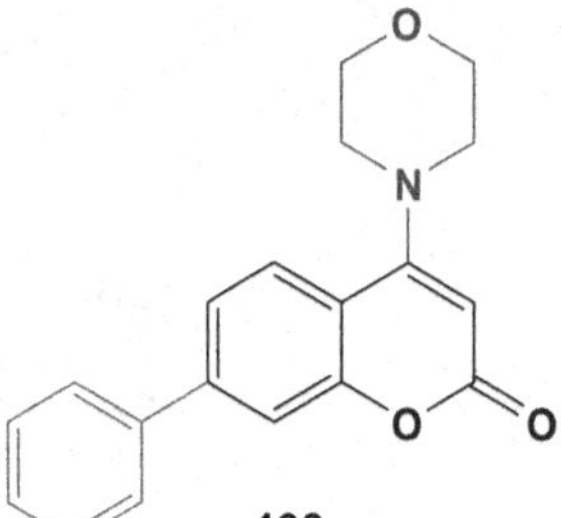

408

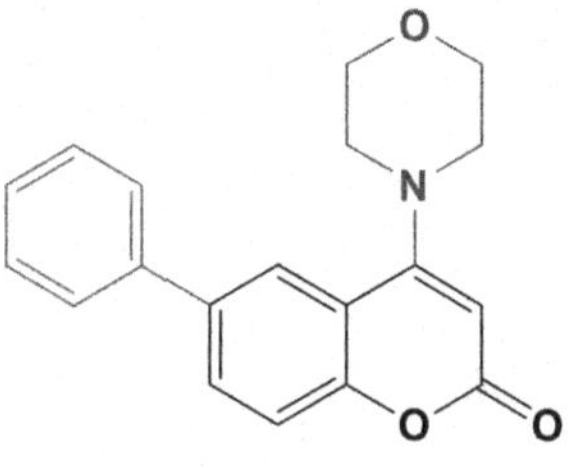

409

9.3 ANTIPSYCHOTIC AGENTS

Antipsychotic agents are utilized for instant treatment of bipolar conditions, to regulate psychotic behavior such as delusions, mania, or hallucinations. These symptoms are usually seen in severe mania or depression. There are two basic classes of antipsychotic agents: *atypical antipsychotic agents* and *typical antipsychotic agents*. Both types of drugs have a tendency to block receptors to the dopamine pathway inside the brain. However, atypical antipsychotic agents also block serotonin receptors. Side effects such as weight gain and difficulty in movement are common with antipsychotics, so it is requisite to check for side effects.

9.3.1 Amine Group

2-(piperidin-4-yl)benzo[d]oxazole [410], 1-tosylpiperazine [413] and 2-(piperidin-4-yl)benzo[d]thiazole [411], (4-fluorophenyl)(piperidin-4-yl)-methanone [412] are amine derivatives that displayed low affinities for all three receptors: 5-HT_{1A}, 5-HT_{2A}, and D_2.[94]

410

411

412

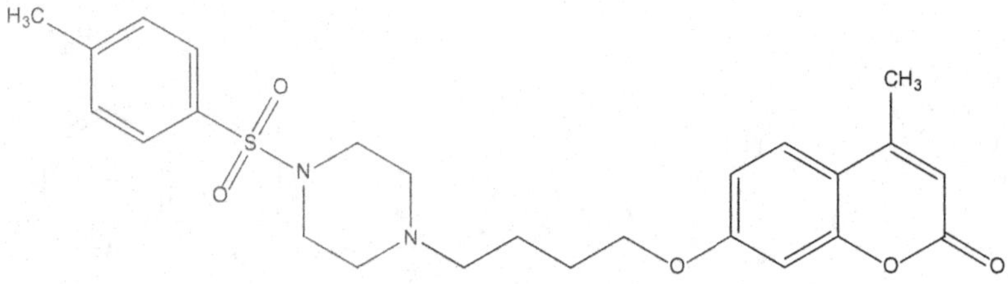

413

An effective affinity for 5-HT$_{1A}$ receptor was shown by amine derivatives such as phenylpiperazines [414–418]. The strongest affinity for 5-HT$_{1A}$ receptors was displayed by the compound 2-methoxyphenylpiperazine [415] (Ki = 0.012 nM).[94]

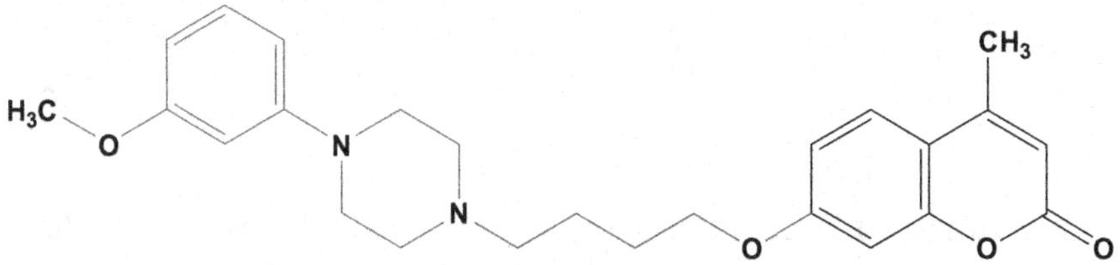

414

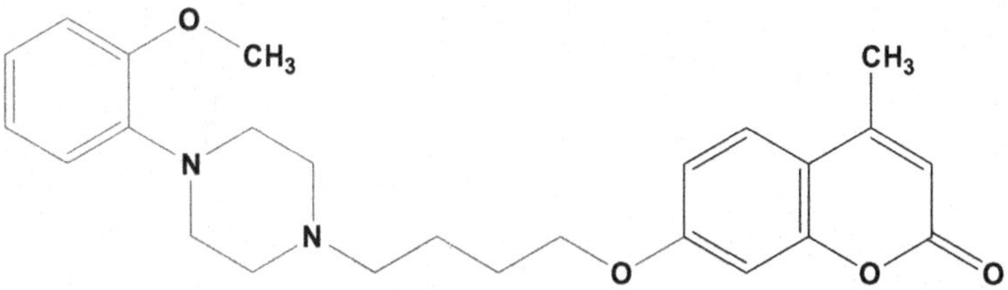

415

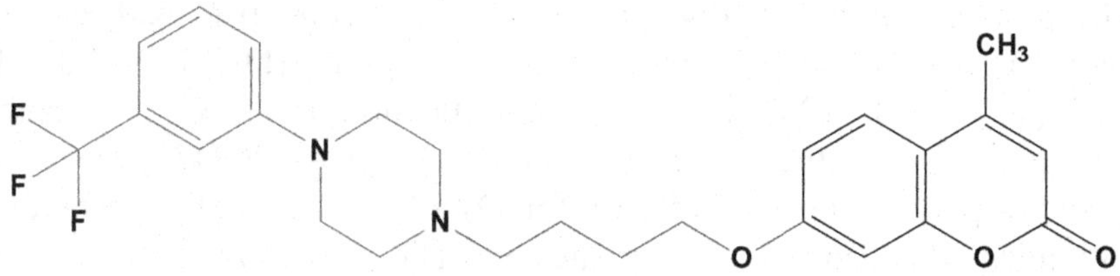

416

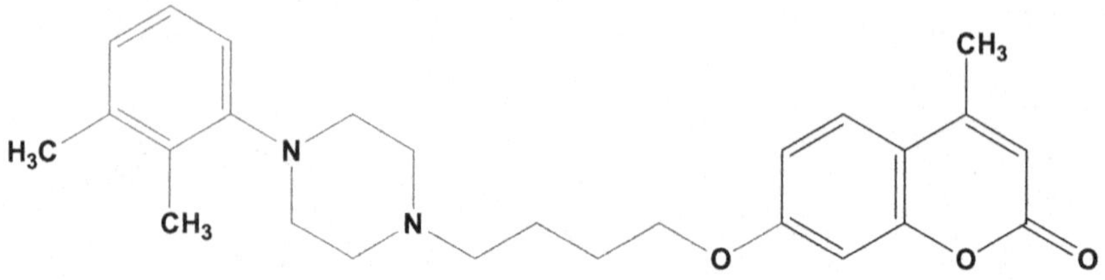

417

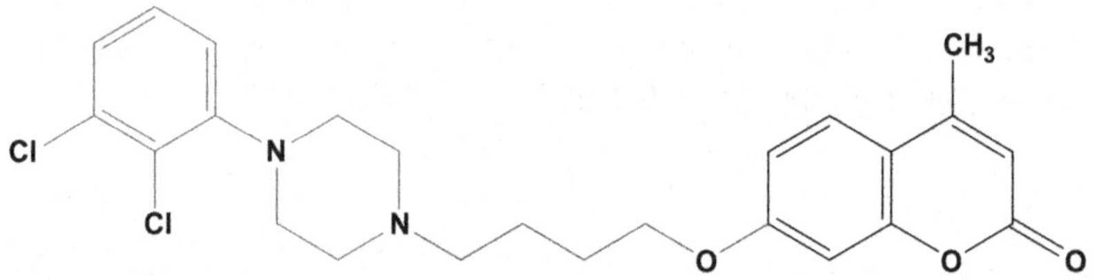

418

The affinity for all three receptors was reduced by the *N*-phenyl group substituted with a pyridine group [**419**].[94]

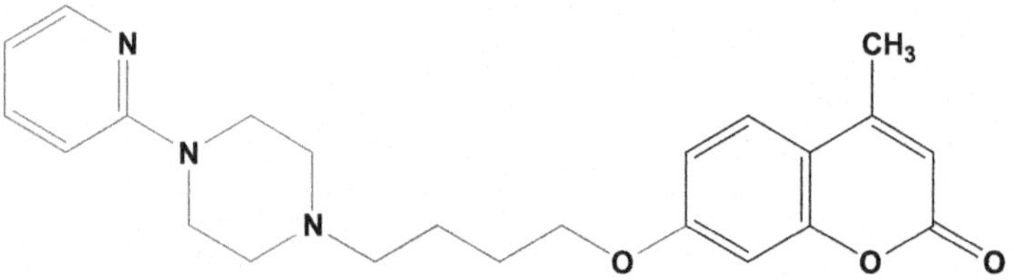

419

9.3.2 Functional Group at C₃ Position

9.3.2.1 Methyl Group

The presence of the methyl group at C-3 of coumarin [**420**] caused increase potency at D_2 and $5\text{-}HT_{2A}$ receptors (D_2, $Ki = 4.0$ nM; $5\text{-}HT_{2A}$, $Ki = 0.3$ nM) without a modification in $5\text{-}HT_{1A}$ receptor activity. However, the derivative with a methoxy group [**421**] as substituents displayed the weak affinity for D_2 as well as $5\text{-}HT_{1A}$ receptors, with adequate affinity for active sites of $5\text{-}HT_{2A}$ receptors.[94]

420

421

9.3.3 Functional group at C_4 Position
9.3.3.1 Alkyl Group
The following order of alkyl substituents in the C-4 position of the coumarin nucleus showed activity for D_2, 5-HT_{1A}, and 5-HT_{2A} receptors: methyl > n-propyl > ethyl > isopropyl > cyclopropyl. Hence methyl substituents at position 4 methyl group of the coumarin ring is important in the variation of serotonergic and dopaminergic activity.[94]

Methyl Group
The replacement of methyl group from the C4 position of coumarin moiety with the H [424–425] and CH_2OH [426–427] caused the decrease in affinities of compounds for D_2, 5-HT_{1A}, and 5-HT_{2A} receptors. Weak affinity for all three receptors was showed by compounds with methoxy group, 2-methoxyphenylpiperazine derivatives [422, 424].[94]

422

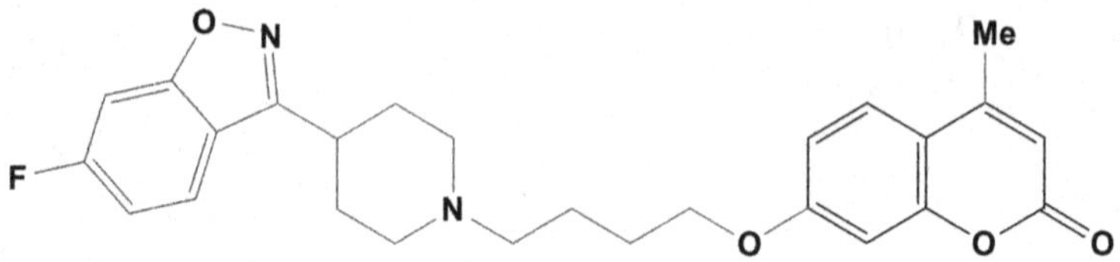

423

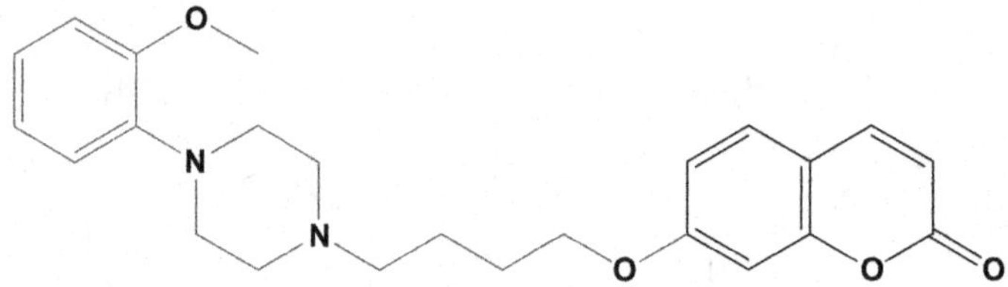

424

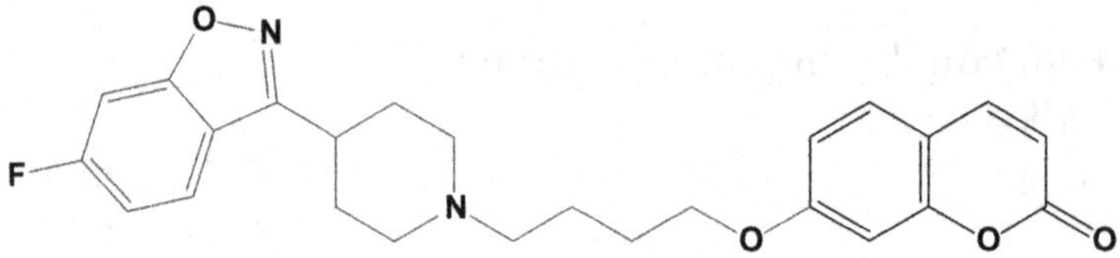

425

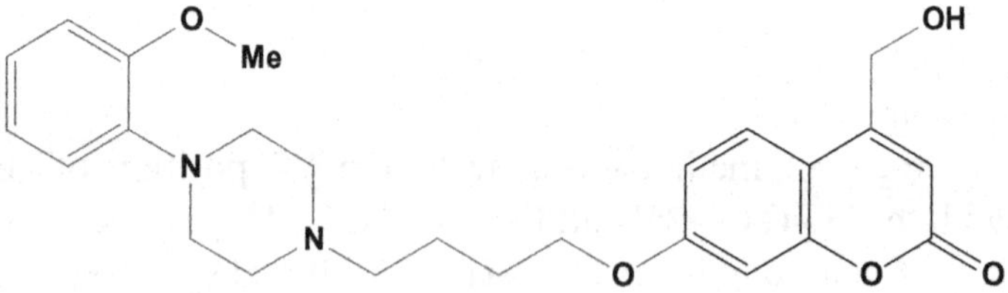

426

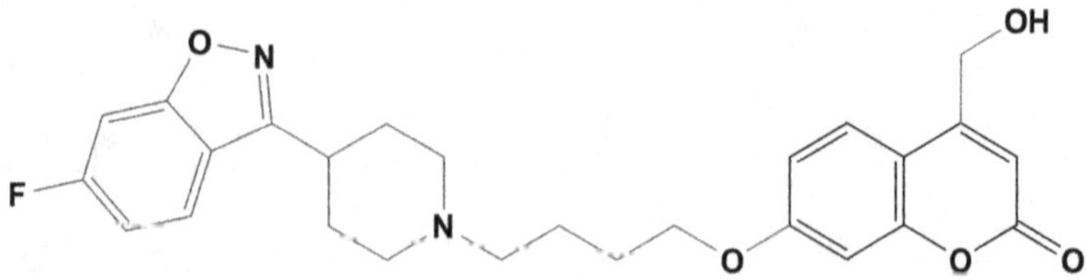

427

9.3.3.2 CF₃ Group (Electron-Withdrawing Group)

Compounds substituted with electron-withdrawing (CF_3) [428–429] showed weak affinity for two $5\text{-}HT_{2A}$ and D_2 receptors; however, there was greater affinity for the $5\text{-}HT_{1A}$ receptor.[94]

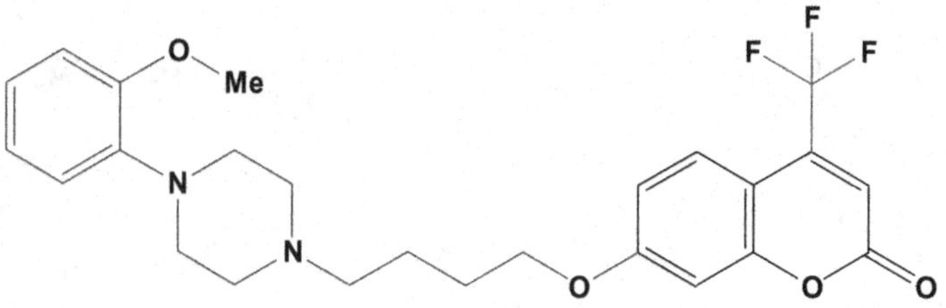

428

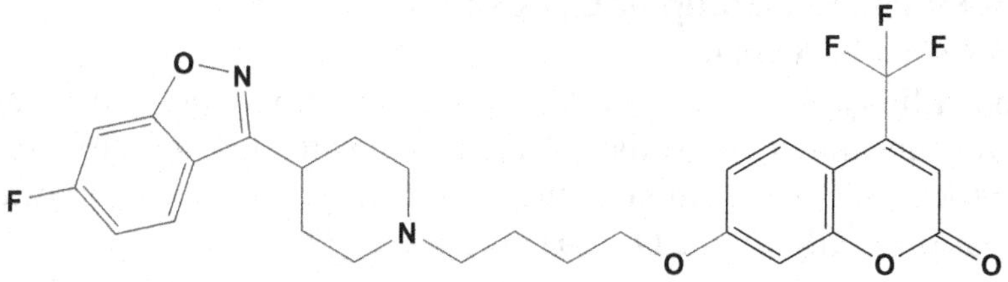

429

9.3.3.3 Phenyl Group

Phenyl-substituted 6-fluorobenzo[d]isoxazol-3-yl)piperidine compound [430] exhibited good affinities for binding site of $5\text{-}HT_{1A}$, $5\text{-}HT_{2A}$, and D_2 ($5\text{-}HT_{1A}$, $K_i = 12.7\,nM$; $5\text{-}HT_{2A}$, $K_i = 9.2\,nM$; D_2, $K_i = 14.8\,nM$).[94]

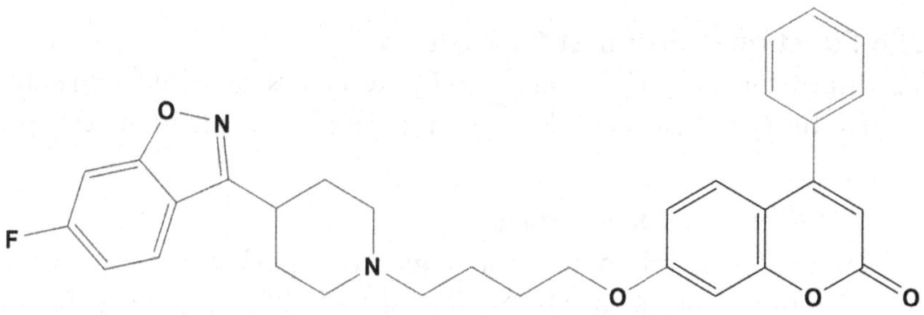

430

9.3.4 Functional Group at C$_5$ Position

9.3.4.1 Methyl Group

A high affinity for 5-HT$_{1A}$ receptors was exhibited through 5-methyl-substituted derivative [431] (Ki = 18.3 nM) and showed reasonable affinity for 5-HT$_{2A}$ and D$_2$ receptors.[94]

431

9.3.5 Functional Group at C$_6$ Position

9.3.5.1 Chlorine Group

Good affinity for 5-HT$_{1A}$ and 5-HT$_{2A}$ receptors was shown by 6-chloro substituted derivative [432] also having low affinity for D$_2$ receptors. The introduction of the substituents at C-5 or C-6 positions decreases the affinity for D$_2$, 5-HT$_{1A}$, and 5-HT$_{2A}$.[94]

432

9.3.6 Functional Group at C$_8$ Position

The affinity for D$_2$, 5-HT$_{1A}$, and 5-HT$_{2A}$ receptors increased with substitution at the C-8 position of the coumarin ring having an amine group.

9.3.6.1 Reduction of Double Bond

Reduction of a double bond in a single bond [433] caused a decrease in affinity of coumarin for D$_2$, 5-HT$_{1A}$, and 5-HT$_{2A}$ receptors (Ki: D$_2$, 515.2 nM; 5-HT$_{1A}$, >10000 nM; 5-HT$_{2A}$, 558.9 nM).[94]

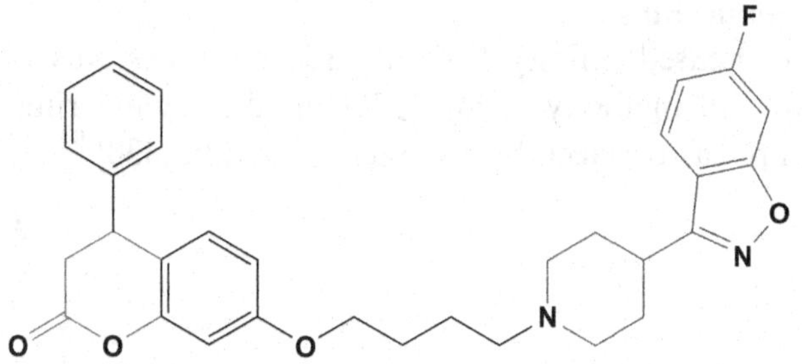

433

9.3.6.2 Methyl Group

Compounds with a methyl group at the eighth position of the couma-rin nucleus with 2-methoxyphenylpiperazine moiety [**434**] showed mod-erate affinities for D_2, 5-HT_{1A}, and 5-HT_{2A} receptors.[94]

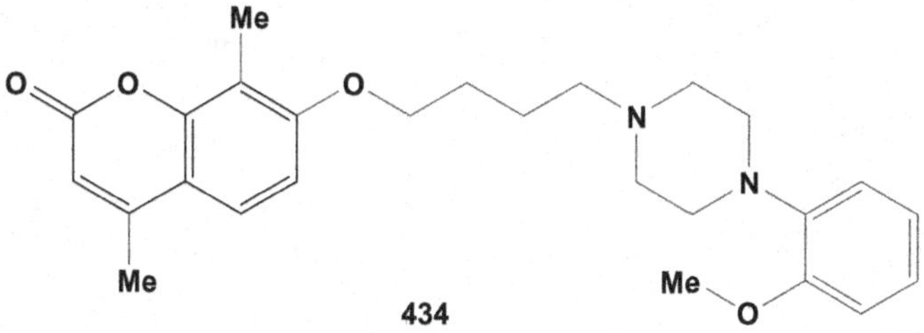

434

9.3.6.3 Chlorine Group

Moderate affinities for D_2, 5-HT_{1A}, and 5-HT_{2A} receptors were shown by Cl moiety at the eighth position of the coumarin compound with a 2-methoxyphenylpiperazine group [**435**].[94]

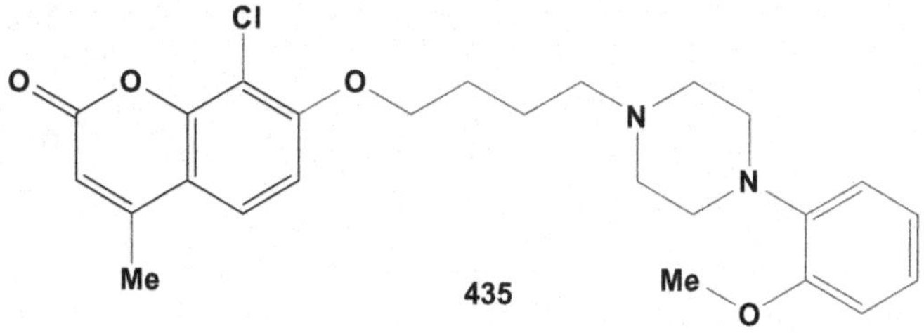

435

9.3.6.4 Aromatic Ring

A drastic decreased affinity for all three receptors was observed for flouro-[436] and methoxy group [437] on the phenyl ring or when a phenyl group was replaced by a heterocyclic ring [438].[94]

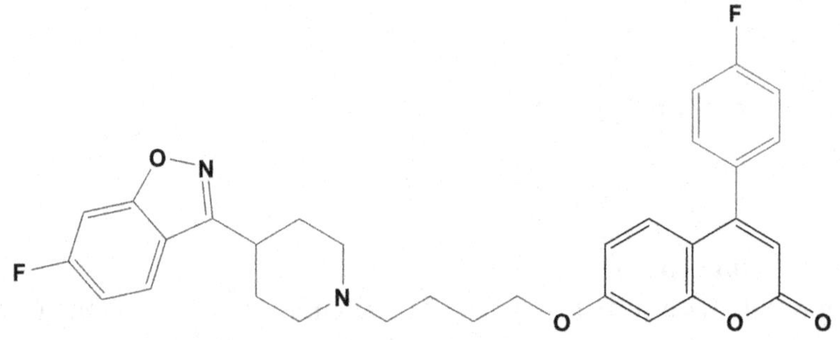

436

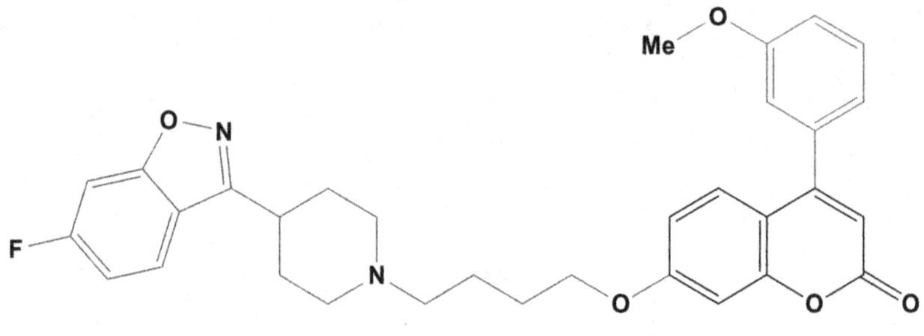

437

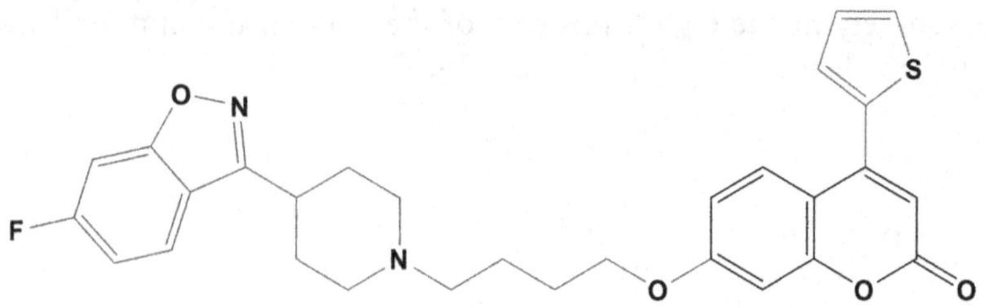

438

9.3.6.5 Alkyl Linker

The length of the linker between the coumarin and the piperidine ring plays an important role for receptor affinity. Chain lengths of three [440] or five [439] carbon atoms significantly reduced D_2, 5-HT_{1A}, and

5-HT$_{2A}$ receptors binding affinity, whereas a four-carbon chain length was the most active.[94]

439

440

9.4 TYROSINASE INHIBITORS

Tyrosinase is also recognized as polyphenol oxidase (PPO). It is a copper-containing enzyme that is broadly spread in animals, plants, and microorganisms. It catalyzes the oxidation reaction in some molecules.[95] Tyrosinase inhibitors are applicable as antifungals, antibacterials, anticonvulsants, food preservatives, and food additives.[96]

9.4.1 Thiosemicarbazides

An important role might played by thiosemicarbzide derivatives [441–443] due to the sulfur atom that forms a complex, which causes loss of catalyting activity of tyrosinase by chelating with the dicopper compound at the binding site.[96]

441

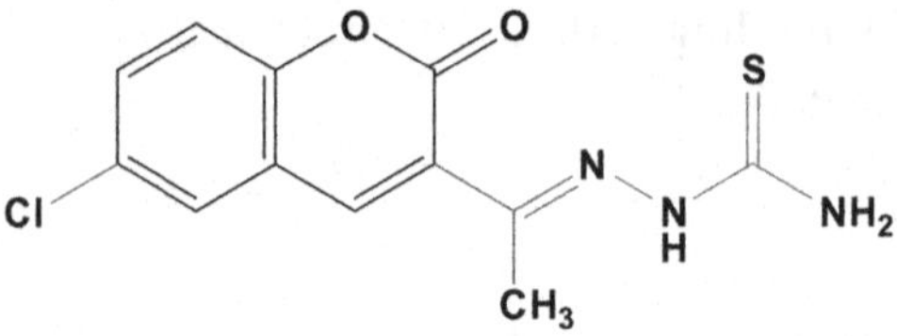

442

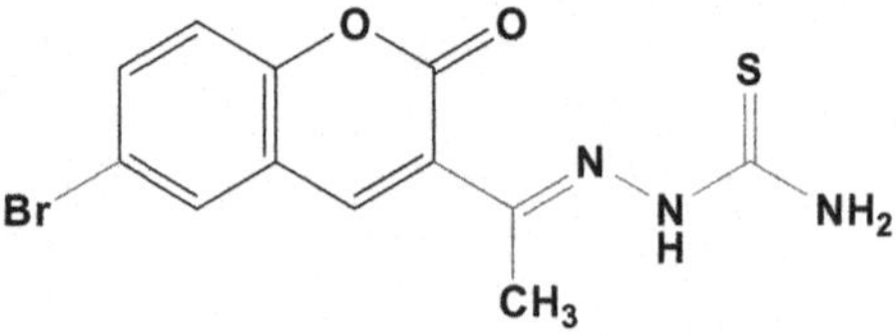

443

9.4.2 Carboxyl Group

Carboxyl substituted compounds [**444**–**445**], might play a very crucial role in interaction between compounds and the active site of tyrosinase.[96]

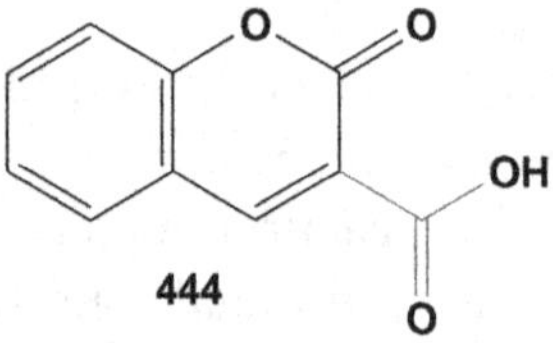

444

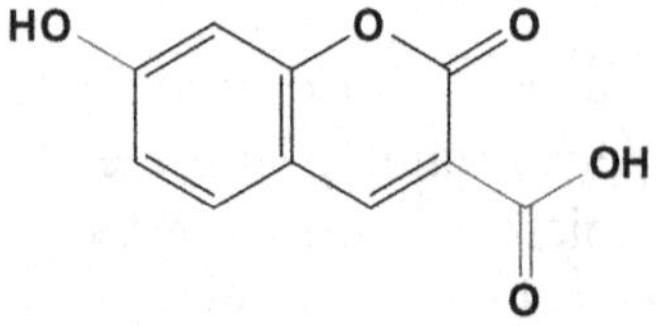

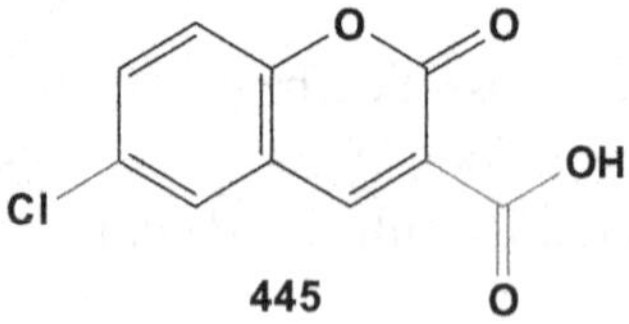

445

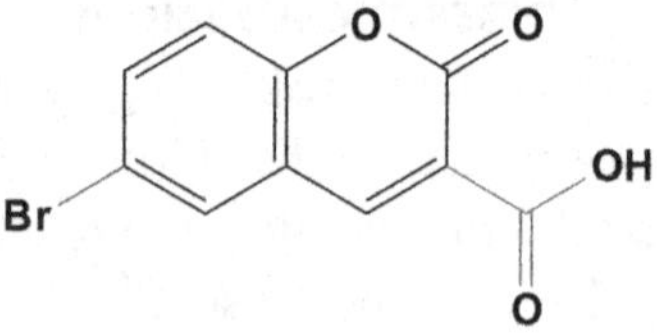

9.4.3 Ester Group

The presence of an ester group in a structure [456–447a–h] can increase the inhibitory activity.[96]

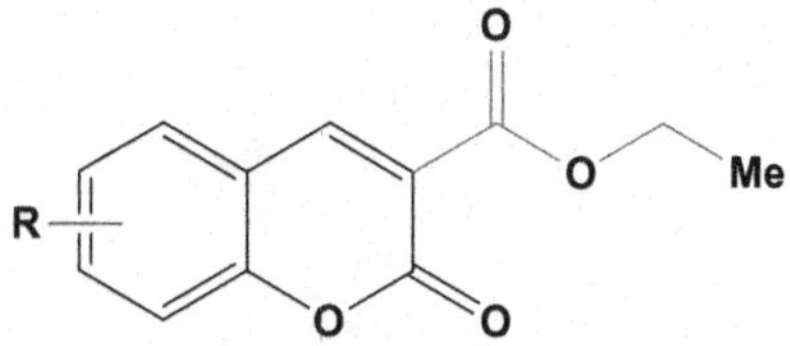

R = H, 6-chloro, 6-bromo, 7-hydroxy

456

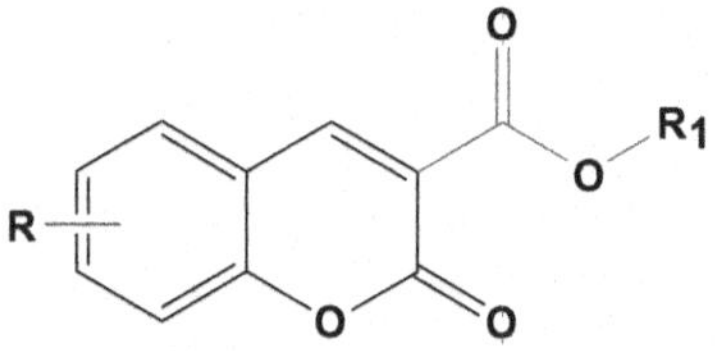

447

447	R	R$_1$
a.	H	Me
b.	7-OH	Me
c.	6-Br	Me
d.	H	Pentyl
e.	7-OH	Pentyl
f.	6-Cl	i-Pr
g.	6-Br	i-Pr
h.	7-OH	i-Pr

REFERENCES

1. Curini M, Cravotto G, Epifano F, Giannone G. *Curr Med Chem* 2006;**13**(2):199–222.

2. Vogel A. Gilberts. *Ann Phys* 1820;**64**:161.

3. Borges F, Roleira F, Milhazes N, Santana L, Uriarte E. *Curr Med Chem* 2005;**12**:887.

4. Dariusz B. *J Chem Research (S)* 1998;468.

5. Aoife L, Richard OK. *Curr Pharm Des* 2004;**10**:3797–811.

6. Feuer G, Kellen JA, Kovacs K. *Oncology* 1976;**33**:35.

7. Kashman Y, Gustafson KR, Fuller RW, Cardellina JH, McMahon JB, Currens MJ, et al. *J Med Chem* 1993;**36**:1110.

8. Shikishima Y, Takaishi Y, Honda G, Ito M, Takfda Y, Kodzhimatov OK, et al. *Chem Pharm Bull* 2001;**49**:877.

9. Gage BF. *Am Soc Hematol Educ Program* 2006;467.

10. Ostrov DA, Hernandez Prada JA, Corsino PE, Finton KA, Le N, Rowe TC. *Antimicrob Agents Chemother* 2007;**51**:3688.

11. Gormley NA, Orphanides G, Meyer A, Cullis PM, Maxwell A. *Biochemistry* 1996;**35**:5083.

12. Fylaktakidou KC, Hadjipavlou-Litina DJ, Litinas KE, Nicolaides DN. *Curr Pharm Des* 2004;**10**:3813.

13. Jing S, Li CJ, Yang JZ, Jie M, Wang C, Jia T, et al. *Fitoterapia* 2014;**96**:138–45.

14. Koneni VS, Gopal RP, Srinivasa RA, Singh S, Manish J, Madhu D. *Bioorg Med Chem Lett* 2012;**22**(9):3115–21.

15. Yogita B, Gulshan B. *Acta Pharm Sin B* 2014;**4**(5):368–75.

16. Powar CB, Daginawala HF. *General microbiology*, vol. II. Himalaya Publishing House; 2003.

17. Bing L, Ramdas P, Ming D, Daniel A, Marjorie HB, Michelle MB, et al. *J Med Chem* 2012;**55**(24):10896–908.

18 Dube RC, Maheswari DK. *General microbiology*. New Delhi: S Chand; 2000.

19. Periers AM, Laurin P, Ferroud D, Haesslein JL, Klich M, Claudine DH, et al. *Bioorg Med Chem Lett* 2000;**10**:161–5.

20. Kanokporn P, Warinthorn C, Pornthep S. *Scientific World J* 2013;11 178649.

21. Zhang BL, Fan CQ, Dong L, Wang FD, Yue JM. *Eur J Med Chem* 2010;**45**:5258–64.

22. Mahantesha B, Vishwanath BJ, Nivedita NB, Sandeep SL, Venkatesh DN. *Eur J Med Chem* 2014;**74**:225–33.

23. Ananthanarayan R. *Textbook of Microbiology*, 2005.

24. Patrick L, Didier F, Laurent S, Michael K, Claudine DH, Pascale M, et al. *Bioorg Med Chem Lett* 1999;**9**:2875–80.

25. Graham DY. *J Gastroenterol Hepatol* 1991;**6**:105–13.

26. Smith HJ. *Introduction to the principles of drug design and action.* 4th ed. CRC Press; 2005; pp. 557–615.

27. Erik DC. *Nat Rev Drug Discov* 2006;**5**:1015–25.

28. Xie L, Takeuchi Y, Mark CL, Andrew TM, Lee KH. *J Med Chem* 2001;**44**:664–71.

29. Tsay SC, Hwu JR, Singha R, Huang WC, Chang YH, Hsu MH, et al. *Eur J Med Chem* 2013;**63**:290–8.

30. Dimitrios V, Leonard HC. *Arthritis Rheum* 2002;**46**(3):585–97.

31. Johan N, Erik De C, Raghunath S, Chang YH, Asish RD, Subhasish KC, et al. *J Med Chem* 2009;**52**:1486–90.

32. Hwu JR, Lin SY, Tsay SC, Raghunath S, Benoy KP, Pieter L, et al. *Phosphorus, Sulfur Silicon Relat Elem* 2011;**186**:1144–52.

33. Hwu JR, Lin SY, Tsay SC, Clercq ED, Pieter L, Johan N. *J Med Chem* 2011;**54**:2114–26.

34. Mahmoud AG, Louis BR. *Clin Microbiol Rev* 1999;**12**(4):501–17.

35. Patrick V, Selene F, Alix TC. *Int J Microbiol* 2012;**2012**:26.

36. Rodrigo SAA, Felipe QSG, Edeltrudes OL, Carlos AS, Josean FT, Luciana S, et al. *Int J Mol Sci* 2013;**14**:1293–309.

37. Bojan S, Maja M, Milan C, Lars G. *Food Chem* 2013;**139**:488–95.

38. Paul EN, Dignani MC, Elias JA. *Clin Microbiol Rev* 1994;479–504.

39. Ambili R. *Int J Basic Clin Pharmacol* 2012;**1**(1):2–12.

40. Rajesh NG, Sharad GJ. *J Exp Clin Med* 2012;**4**(3):165–9.

41. Alfonso M, Claudiu TS. *Bioorg Med Chem Lett* 2010;**20**:4511–14.

42. Tehsina D, Claire R, Christina W, Suresh A, Shigetoshi K, Dora CM. *Bioorg Med Chem Lett* 2011;**21**:5770–3.

43. Fabrizio C, Alfonso M, Andrea S, Claudiu TS. *Bioorg Med Chem Lett* 2012;**20**:2266–73.

44. Nadia T, Alfonso M, Paul CM, Yuanmei L, Andrea S, Shoukat D, et al. *J Med Chem* 2011;**54**:8271–7.

45. Elzbieta B, Magdalena M, Lorenz IP, Mayer P, Renata AK, Piotr P, et al. *Inorg Chem* 2006;**45**:9688–95.

46. Wang ZC, Qin YJ, Wang PF, Yang YA, Wen Q, Zhang X, et al. *Eur J Med Chem* 2013;**66**:1–11.

47. Tamer N, Samir B, Mahmoud Y. *Eur J Med Chem* 2014;**76**:539–48.

48. Gaelle LB, Christine R, Peyrat JF, Brion JD, Alami M, Marsaud V, et al. *J Med Chem* 2007;**50**:6189–200.

49. Wenjuan Z, Zhi L, Meng Z, Feng W, Xueyan H, Hao L, et al. *Bioorg Med Chem Lett* 2014;**24**:799–807.

50. Bandi Y, Uma DH, Rajashaker B, Lingaiah N, Ganesh KC, Sujitha P, et al. *Eur J Med Chem* 2014;**79**:260–5.

51. Wang X, Kyoko NG, Kenneth FB, Don MJ, Lin YL, Wu TS, et al. *J Med Chem* 2006;**49**:5631–4.

52. Maria ER, Dominick M, Ramiro V, Monica V, Sven M, Jan J, et al. *Bioorg Med Chem* 2009;**17**:6547–59.

53. Shulin H, Vicki Z, Shifeng P, Yi L, Michael H, Daniel M, et al. *Bioorg Med Chem Lett* 2005;**15**:5467–73.

54. El-Gamal MI, Oh CH. *Eur J Med Chem* 2014;**84**:68–76.

55. Alfonso M, Andrea S, Claudiu TS. *Bioorg Med Chem Lett* 2010;**20**:7255–8.

56. Alexel V, Frank K. *Neuroscientist* 2007;**13**:28.

57. Mark WI, Blaise MC, Arturas V, Guangyu F, Laura C, Graham LC, et al. *Neurochem Int* 2012;**61**:593–600.

58. Kang SY, Young CK. *Arch Pharm Res* 2007;**30**(11):1368–73.

59. Brookmeyer R, Gray S, Kawas C. *Am J Public Health* 1998;**88**(9):1337–42.

60. Shinsuke M, Mitsuo M. *Bioorg Med Chem* 2012;**20**:784–8.

61. Angela R, Alessandra B, Piero V, Maurizio R, Andrea C, Vincenza A, et al. *J Med Chem* 1998;**41**:3976–86.

62. Masoumeh A, Mehdi K, Alireza F, Hamid N, Alireza M, Amirhossein S, et al. *Bioorg Med Chem* 2012;**20**:7214–22.

63. Lorna P, Andrea C, Federica B, Alessandra B, Silvia G, Stefano R, et al. *J Med Chem* 2007;**50**:4250–4.

64. Cedrik G, Taisuke T, Nicolas P, Younes L, Roselyne R, Gaetan H, et al. *J Med Chem* 2006;**49**:4275–85.

65. Kang SY, Lee KY, Sung SH, Park MJ, Kim YC. *J Nat Prod* 2001;**64**:683–5.

66. Maria JM, Carmen T, Yunierkis PC, Eugenio U, Lourdes S, Dolores V. *J Med Chem* 2011;**54**(20):7127–37.

67. Serra S, Ferino G, Matos MJ, Vázquez RS, Delogu G, Vina D, et al. *Bioorg Med Chem Lett* 2012;**22**(1):258–61.

68. Maria JM, Santiago V, Rosa MGF, Eugenio U, Lourdes S, Carol F, et al. *Eur J Med Chem* 2013;**63**:151–61.

69. Carmela G, Marco C, Francesco L, Peter W, Carrupt PA, Cosimo A, et al. *J Med Chem* 2000;**43**:4747–58.

70. Leonardo P, Marco C, Orazio N, Giancarlo G, Mario DB, Ramon SO, et al. *Eur J Med Chem* 2013;**70**:723–39.

71. Cecilia M, Peder S, Clas S. *Eur J Med Chem* 2014;**73**:177–86.

72. Matthias DM, Sonja H, Christa EM, Michael G. *Bioorg Med Chem* 2014;**22**:1916–28.

73. Omaima MA, Kamelia MA, Hamed IA, Mohamed MA, Rasha ZB. *J Med Chem* 2012;**55**(23):10424–36.

74. Angelo C, Antonio C, Stefano C, Marco B, Barbara C, Carmela G, et al. *Bioorg Med Chem Lett* 2002;**12**:3551–5.

75. Cheng JF, Chen M, David W, Sovouthy T, Thomas A, Hirotaka K, et al. *Bioorg Med Chem Lett* 2004;**14**:2411–15.

76. Erich LG, Christine B, Nathalie C, Chan CC, Daniel D, Yves D, et al. *Bioorg Med Chem Lett* 2006;**16**:2528–31.

77. Silke BB, Cedrick M, Wolfram H, Heike W. *Eur J Med Chem* 2009;**44**:1838–52.

78. Rangappa SK, Kallappa MH, Ramya VS, Mallinath HH. *Eur J Med Chem* 2010;**45**:2597–605.

79. Amanda SR, Gabriel NV, Sergio HF, Maria YR, Maximiliano IB, Samuel ES. *Eur J Med Chem* 2014;**77**:400−8.

80. Judith S, Igor B, Gerhard FE, Brigitte K, Steffen H. *Eur J Pharmacol* 2011;**668**:57−64.

81. Phuong TT, Tran MH, Tran MN, Do TH, Byung SM, Seung JK, et al. *Phytother Res* 2010;**24**:101−6.

82. Lin HC, Tsai SH, Chen CS, Chang YC, Lee CM, Lai ZY, et al. *Biochem Pharmacol* 2008;**75**:1416−25.

83. Zhang HY, Wang LF. *J Mol Struc-Theochem* 2004;**673**:199−202.

84. Milan M, Mirjana M, Desanka B, Sanja M, Neda N, Vladimir M, et al. *Int J Mol Sci* 2011;**12**:2822−41.

85. Som DS, Hament KR, Shilpa C, Rakesh KS. *Bio Metals* 2005;**18**:143−54.

86. Francisco JMM, Rodrigo SRH, Ana LPC, Manuel VG, Maria TSM, Daniel JC, et al. *Molecules* 2012;**17**:14882−98.

87. Sarah J, Karam C, Abha K, Prabhjot S, Nivedita P, Bhavna G, et al. *Bioorg Chem* 2012;**40**:131−6.

88. Raphae F, Caroline C, Severine R, Johan W, Bernard M, Lionel P. *Bioorg Med Chem Lett* 2006;**16**:2017−21.

89. Giorgio R, Mario DB, Giancarlo G, Daniela P, Giuliana L, Debora B, et al. *J Med Chem* 2007;**50**:2886−95.

90. Yang YJ, Lee HJ, Lee BK, Lim SC, Lee CK, Lee MK. *Fitoterapia* 2010;**81**:497.

91. Omaima MA, Kamelia MA, Hamed IA, Timothy JM, Rasha ZB. *Neurochem Int* 2011;**59**:906−12.

92. Saleh KI, Jasim MAAR, Christopher JB, Michael JA, Murray NR. *Eur J Med Chem* 2012;**57**:85−101.

93. Sara LP, Sonsoles RA, Julia B, Celine C, Bernard TG, Ian RH, et al. *Bioorg Med Chem Lett* 2010;**20**:3649−53.

94. Yin C, Songlin W, Xiangqing X, Xin L, Minquan Y, Song Z, et al. *J Med Chem* 2013;**56**(11):4671−90.

95. Liu JB, Yi W, Wan YQ, Ma L, Song HC. *Bioorg Med Chem* 2008;**16**(3):1096−102.

96. Jinbing L, Fengyan W, Lingjuan C, Liangzhong Z, Zibing Z, Min W, et al. *Food Chem* 2012;**135**:2872−8.